T0350290

Enterprise Wireless Local Area Network Architectures and Technologies

Data Communication Series

For more information on this series please visit: https://www.routledge.com/
Data-Communication-Series/book-series/DCSHW

Enterprise Wireless Local Area Network Architectures and Technologies

Rihai Wu, Xun Yang, Xia Zhou,
and Yibo Wang

CRC Press
Taylor & Francis Group
Boca Raton London New York

CRC Press is an imprint of the
Taylor & Francis Group, an **informa** business

人民邮电出版社
POSTS & TELECOM PRESS

First edition published 2021
by CRC Press
6000 Broken Sound Parkway NW, Suite 300, Boca Raton, FL 33487-2742

and by CRC Press
2 Park Square, Milton Park, Abingdon, Oxon, OX14 4RN

© 2021 Rihai Wu, Xun Yang, Xia Zhou and Yibo Wang
Translated by Xiangjia Ji

CRC Press is an imprint of Taylor & Francis Group, an Informa business

No claim to original U.S. Government works

**Visit the Taylor & Francis Web site at
http://www.taylorandfrancis.com**

**and the CRC Press Web site at
http://www.crcpress.com**

English Version by permission of Posts and Telecom Press Co., Ltd.

Library of Congress Cataloging-in-Publication Data
Names: Wu, Rihai, author.
Title: Enterprise wireless local area network architectures and technologies / Rihai Wu, Xun Yang,
Xia Zhou, Yibo Wang.
Description: First edition. | Boca Raton, FL : CRC Press, 2021. | Includes
bibliographical references. | Summary: "This book has been written with the support of Huawei's large
accumulation of technical knowledge and experience in the WLAN field, as well as its understanding
of customer service requirements. The book first covers service challenges facing enterprise wireless
networks, along with detailing the latest evolution of Wi-Fi standards, air interface performance, and
methods for improving user experience in enterprise scenarios. Furthermore, it illustrates typical networking, planning, and scenario-specific design for enterprise WLANs, and provides readers with a
comprehensive understanding of enterprise WLAN planning, design, and technical implementation,
as well as suggestions for deployment. This is a practical and easy-to-understand guide to WLAN
design, and is written for WLAN technical support and planning engineers, network administrators,
and enthusiasts of network technology"—Provided by publisher.
Identifiers: LCCN 2020050041 (print) | LCCN 2020050042 (ebook) |
ISBN 9780367695750 (hardcover) | ISBN 9781003143659 (ebook)
Subjects: LCSH: Wireless LANs.
Classification: LCC TK5105.78 .W83 2021 (print) | LCC TK5105.78 (ebook) | DDC 004.6/8—dc23
LC record available at https://lccn.loc.gov/2020050041
LC ebook record available at https://lccn.loc.gov/2020050042

ISBN: 978-0-367-69575-0 (hbk)
ISBN: 978-0-367-69875-1 (pbk)

Typeset in Minion Pro
by codeMantra

Contents

Summary

THIS BOOK HAS BEEN written with the support of Huawei's large accumulation of technical knowledge and experience in the WLAN field, as well as its understanding of customer service requirements. The book first covers service challenges facing enterprise wireless networks, along with detailing the latest evolution of Wi-Fi standards, air interface performance, and methods for improving user experience in enterprise scenarios. Furthermore, it illustrates typical networking, planning, and scenario-specific design for enterprise WLANs, and provides readers with a comprehensive understanding of enterprise WLAN planning, design, and technical implementation, as well as suggestions for deployment.

This is a practical and easy-to-understand guide to WLAN design, and is written for WLAN technical support and planning engineers, network administrators, and enthusiasts of network technology.

Introduction

THIS BOOK HAS BEEN published to coincide with the 21st anniversary of Wireless Fidelity (Wi-Fi) technology, following the founding of the Wireless Ethernet Compatibility Alliance (WECA, renamed as Wi-Fi Alliance in 2002) and the 802.11a/b release in 1999. Wi-Fi has fluctuated in its popularity over the years, but grew substantially through rapid iterations of its technology, dominating the short-range wireless access field.

Wi-Fi has been applied to enterprise networks for more than 10 years. Wireless Local Area Network (WLAN) using Wi-Fi was originally only a supplement to wired networks. Nowadays, it has been integrated with wired networks, with Bring Your Own Device (BYOD) and wireless office in wide use. Wi-Fi has become essential to enterprise networks and helps enterprises increase productivity and innovation.

The increasing significance of enterprise WLANs has raised the increasingly higher requirements of customers. Huawei's end-to-end (E2E) authentication and security services, highly reliable network services, and high-quality products provide customers with excellent wireless access, seamless roaming, and accessible ultra-broadband anytime, anywhere.

This book leverages Huawei's extensive technical knowledge and experience in the WLAN field, as well as its understanding of customer service requirements. The book first covers service challenges facing enterprise WLANs and introduces new Wi-Fi standards, as well as the solutions to improving air interface performance and user experience, wireless security defense measures, and Internet of Things (IoT) integration solutions in enterprise scenarios. Furthermore, it illustrates typical networking, planning, and scenario-specific design for enterprise WLANs and provides readers with a comprehensive understanding of enterprise WLAN solutions as well as suggestions for deployment. It is helpful for ICT professionals, such as network engineers, who want to plan, design, and deploy

WLANs, and understand WLAN architecture. This book is also an excellent reference for enthusiasts of network technology and students looking to learn about new WLAN technologies.

How Is the Book Organized

This book consists of 10 chapters. Synopses of chapters follows below.

Chapter 1: Enterprise WLAN Overview

This chapter first details the emergence and evolution of Wi-Fi technology, as well as the development phases of enterprise WLAN. It then covers the trends and challenges facing enterprise WLANs, and introduces the next-generation enterprise WLAN solution that addresses these challenges.

Chapter 2: WLAN Technology Basics

IEEE 802.11 standards formulated by the IEEE 802.11 working group are currently used as WLAN standards.

The IEEE established the 802.11 working group in 1990 to standardize WLAN, and after decades of development, 802.11 has evolved into a suite of standards. This chapter first introduces the basic concepts of wireless communications and key WLAN technologies, followed by the evolution and comparison of 802.11 standards. Finally, it summarizes the next-generation 802.11ax standard and key technologies at the physical and MAC layers.

Chapter 3: Air Interface Performance and User Experience Improvement

Wi-Fi technology holds its strong position, mainly due to its ultra-high bandwidth. The bandwidth increased from 11 Mbps in the earliest 802.11b standard to 10 Gbps in the latest 802.11ax standard, greatly improving user experience. However, when there are a large number of users accessing a WLAN, bandwidth per user decreases significantly, especially in high-density areas such as offices. This chapter analyzes the key factors that affect WLAN performance and provides technical approaches for improving air interface performance and user experience.

Chapter 4: WLAN Security and Defense

WLAN uses radio waves in place of network cables to transmit data, allowing for simple infrastructure construction. However, the nature of wireless transmission makes it crucial to guarantee WLAN security. This chapter covers WLAN security threats and mechanisms, such as common access authentication modes, and wireless attack detection and countermeasure.

Chapter 5: WLAN and IoT Convergence

This chapter first introduces several common wireless communications technologies in the IoT, then discusses the feasibility of converged deployment of short-range wireless communications technology used by Wi-Fi networks and IoT, and finally illustrates several typical application scenarios of IoT access points (APs).

Chapter 6: WLAN Positioning Technologies

This chapter covers the principles of common wireless positioning technologies to enable you to quickly master the theoretical basis and better understand the content in subsequent chapters. This chapter then illustrates how to calculate the suitable locations of various short-range wireless communication systems by using the positioning technologies. Last, this chapter covers how to build a positioning system based on actual requirements.

Chapter 7: Enterprise WLAN Networking Design

Various enterprises have different WLAN requirements. As WLANs are widely used in enterprise networks, building a WLAN that meets service requirements is important for enterprises. Before building a solid network, you need to design a good architecture and select the appropriate networking mode. This chapter covers the WLAN networking design and typical networking solutions.

Chapter 8: Enterprise WLAN Planning and Design

Adequate network planning and design can meet the requirements for wide signal coverage, conflict prevention, and large network capacity. This chapter covers the enterprise WLAN planning and design methods based on network construction.

Chapter 9: Scenario-Based Enterprise WLAN Design

This chapter provides scenario analysis and network planning design for typical scenarios, to help you get a grasp of WLAN network planning design methods and apply them to actual projects.

Chapter 10: Enterprise WLAN O&M

Network O&M is an indispensable routine network maintenance task for campus network administrators to ensure normal and stable network running. This chapter focuses on routine monitoring, network inspection, device upgrade, and troubleshooting during network O&M.

Icons Used in This Book

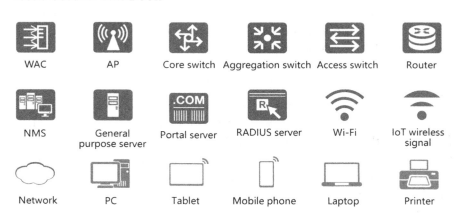

| WAC | AP | Core switch | Aggregation switch | Access switch | Router |

| NMS | General purpose server | Portal server | RADIUS server | Wi-Fi | IoT wireless signal |

| Network | PC | Tablet | Mobile phone | Laptop | Printer |

Acknowledgments

THIS BOOK HAS BEEN jointly written by the Information Digitalization and Experience Assurance (IDEA) Department, DC, and the Data Communication Architecture & Design Department of Huawei Technologies Co., Ltd. During the writing of the book, high-level management from Huawei's Data Communication Product Line provided much guidance, support, and encouragement. Here, we would like to express our sincere gratitude for their support.

The following is a list of participants involved in the preparation and technical review of this book.

Editorial board: Rihai Wu, Yang Xun, Mingfu Ye, Bai Xiaofei, Zhou Xia, Wang Yibo, Ding Yuquan, Yu Juzheng, Ma Zhi, Wang Junhui, Zhou Qi, Ji Xiang, Zhou Xiao, Wang Haitao, Cui Jianlei, Pan Chun, Chen Jian, Wang Mingyue, Gu Weiwei, Ma Jiabin, Li Hongxuan, Luo Xiaoyan, Cai Yajie, Huang Guogang, Jia, Gan Ming, Yu Jian, Guo Yuchen, Li Yunbo

Technical reviewers: Guo Jun, Rihai Wu, Zhou Xia, Wang Yibo, Yang Xun, Mingfu Ye, Bai Xiaofei

Translators: Ji Xiangjia, Hu Ranran, Han Qiang, Chen Xiexia, Liu Cen, Mu Li, Wang Lili, Feng Qiangqiang, Huang Yonggen, George Fahy, and Evan Reeves

While the writers and reviewers of this book have many years of experience in WLAN and have made every effort to ensure accuracy, it may be possible that minor errors have been included due to time limitations. We would like to express our heartfelt gratitude to the readers for their unremitting efforts in reviewing this book.

Authors

Mr. Rihai Wu is Chief Architect of Huawei's campus network WLAN solution with 16 years of experience in wireless communications product design and a wealth of expertise in network design and product development. He previously served as a designer and developer of products for Wideband Code Division Multiple Access (WCDMA), LTE indoor small cells, and WLAN.

Mr. Xun Yang is a WLAN standard expert from Huawei. He has nine years of experience in formulating WLAN standards, and previously served as 802.11ac Secretary, 802.11ah PHY Ad-hoc Co-chair, and 802.11ax MU Ad Hoc Sub Group Co-chair. Mr. Yang oversees technical research, the promotion of standards, and industrialization in the WLAN field, and has filed more than 100 patents.

Ms. Xia Zhou is a documentation engineer of Huawei's campus network WLAN solution. She has 10 years of experience in creating documents for campus network products. Ms. Zhou was previously in charge of writing manuals for Huawei data center switches, WLAN products, and campus network solutions. She is also the author of *Campus Network Solution Deployment Guide* and was a co-sponsor of technical sessions such as WLAN from Basics to Proficiency.

Mr. Yibo Wang is a documentation engineer of Huawei's campus network WLAN solution. He has nine years of experience in creating documents for campus network products. Mr. Wang was previously in charge of writing manuals for Huawei switches, WLAN products, and routers. He was also a co-sponsor of technical sessions such as WLAN from Basics to Proficiency and HCIA-WLAN certification training courses.

CHAPTER 1

Enterprise WLAN Overview

T HIS CHAPTER FIRST DETAILS the emergence and evolution of Wi-Fi technology, as well as the development phases of enterprise wireless local area network (WLAN). It then covers the trends and challenges facing enterprise WLANs, and introduces the next-generation enterprise WLAN solution that addresses these challenges.

1.1 HISTORY OF ENTERPRISE WLAN DEVELOPMENT

Wi-Fi technology is developed to eliminate the limitation of cabling on wired networks. When providing services for customers such as department stores and supermarkets, NCR (a cash register manufacturer) found that it was time- and labor-consuming to reroute cash registers every time the store layout changed. In 1988, together with AT&T and Lucent, NCR launched the WaveLAN wireless network technology solution to integrate wireless access into cash registers. This technology is recognized as the predecessor of Wi-Fi. The development of WaveLAN technology promoted the establishment of the 802.11 WLAN Working Committee in 1990 by the Institute of Electrical and Electronics Engineers (IEEE) 802 Local Area Network (LAN)/Metropolitan Area Network (MAN) Standards Committee. The 802.11 WLAN Working Committee subsequently started to formulate WLAN technical standards.

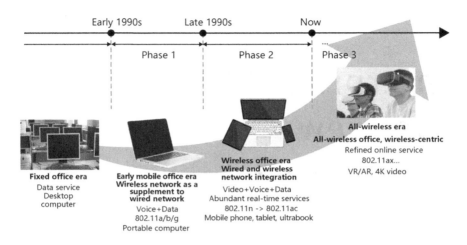

FIGURE 1.1   Enterprise WLAN development history.

Figure 1.1 shows the development of enterprise WLAN in accordance with the WLAN technology evolution, enterprise WLAN architecture, and protocols.

1.1.1 Phase 1: Initial Mobile Office Era — Wireless Networks As a Supplement to Wired Networks

The WaveLAN technology is considered as the prototype of enterprise WLAN. Early Wi-Fi technologies were mainly applied to Internet of Things (IoT) devices such as wireless cash registers. However, with the release of 802.11a/b/g standards, wireless connections have become increasingly advantageous, and the costs of terminal modules have also been decreasing rapidly. Enterprises and consumers are beginning to see the potential of Wi-Fi technologies, and wireless hotspots begin to appear in cafeterias, airports, and hotels. The name Wi-Fi was also created during this period. It is the trademark of the Wi-Fi Alliance. The original goal of the alliance was to promote the formulation of the 802.11b standard and the compatibility certification of Wi-Fi products worldwide. With the evolution of standards and the popularization of standard-compliant products, people tend to equate Wi-Fi with the 802.11 standard. It should be noted that the 802.11 standard is one of many WLAN technologies, and yet it has become a mainstream standard in the industry. When WLAN is mentioned, it usually is a WLAN using the Wi-Fi technology. The first phase of enterprise WLAN application eliminated the limitation of wired access to enable devices to move around freely within a certain range; that is, WLAN extends wired networks with the utilization of wireless networks. In this phase, WLANs do not have specific requirements on security, capacity, and roaming capabilities. APs are still single access points used for wireless coverage in single-point networking. Generally, an AP using a single access point architecture is called a fat AP.

1.1.2 Phase 2: Wireless Office Era — Integration of Wired and Wireless Networks

With the increasing popularity of wireless devices, enterprise WLANs have evolved from supplementing wired networks to being as essential as wired networks. In this phase, an enterprise WLAN, as a part of the enterprise network, needs to meet BYOD requirements and provide network access for enterprise guests.

A single cafeteria usually has dozens of access users, whereas an enterprise may include hundreds or thousands of access users. In this case, a single AP cannot meet the requirements of large-scale WLAN construction, giving rise to the emergence of a new WLAN architecture: WAC+Fit AP. A centralized wireless access controller (WAC) has been introduced into networks to implement centralized channel management, unified configuration, and network-wide roaming and authentication over Control and Provisioning of Wireless Access Points (CAPWAP), as illustrated in Figure 1.2.

In addition, numerous large-bandwidth services, such as video and voice, are required in enterprise office scenarios, thereby imposing higher

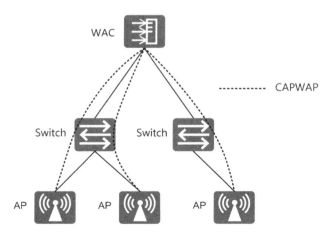

FIGURE 1.2   WAC+Fit AP architecture.

bandwidth requirements on enterprise WLANs. Since 2012, the 802.11ac standard has become mature and implemented many improvements in the operating frequency bands, channel bandwidths, and modulation and coding schemes (MCSs). Compared with earlier 802.11 standards, the 802.11ac standard includes higher traffic volumes and less interference, and it allows more users to access networks, thereby leading WLANs into the gigabit era. This technology improvement represents a solid foundation for enterprise wireless office. In this context, APs need to support the new 802.11 standard and also provide stronger service processing capabilities. Enterprise and home APs are evolving separately.

As the infrastructure of enterprise networks, enterprise APs require a longer service life and a more solid structure design, as well as more powerful processing capabilities and more efficient air interface scheduling algorithms. This will enable APs to meet high concurrency and large capacity access requirements. Table 1.1 lists the differences between enterprise APs and home APs.

TABLE 1.1   Differences between Enterprise APs and Home APs

Item	Home AP Features	Enterprise AP Features
Chip	The consumer-grade chip has a weak processing capability, supporting access of 5–10 users and providing 100 Mbps bandwidth for the entire system	The high-performance industrial-grade chip has a powerful processing capability, supporting access of at least 100 users and providing a bandwidth greater than 1 Gbps for the entire system
Printed circuit board (PCB)	Most PCBs use two layers, whereas some use four. Digital noises are easily generated to interfere with RF signals	PCBs have four to eight layers, facilitating reduction in noise interference and improvement in the quality of RF signals
2.4 GHz RF calibration	Not calibrated. It is possible that good signal strength is displayed while the communication quality for some terminals is poor	Strictly tested and calibrated
Antenna	APs are usually installed on desks or the ground. Generally, external antennas are used	To facilitate installation and achieve better coverage, built-in antennas are used with smart antennas and small-angle directional antennas to meet the requirements of various application scenarios

(Continued)

TABLE 1.1 (*Continued*)   Differences between Enterprise APs and Home Aps

Item	Home AP Features	Enterprise AP Features
Power over Ethernet (PoE)	Not supported. An additional power adapter is required	PoE, PoE+, and PoE++ power supply standards are supported
Lifespan	Less than five years	5–8 years
Wireless standards	Only 802.11ac Wave 1 and earlier standards are supported	802.11ac Wave 2 is widely supported. The latest APs support the 802.11ax standard
Security	Basic security functions, such as host and guest service set identifier (SSID) isolation and terminal Media Access Control (MAC) address whitelist, are supported	Features higher security. Generally, more than 16 SSIDs are supported. Various SSIDs can be allocated to different users. Multiple SSIDs and virtual local area networks (VLANs) can be divided to provide independent subnets as well as different authentication modes and access policies, implementing end-to-end data security isolation from wireless to wired networks
QoS	Internet access time limit and network speed control can be configured for different terminals	The Internet access time limit and network speed control can be configured for terminals as well as applications based on SSIDs, and more refined access control is performed by using uniform resource locators (URLs)
Continuous networking capability	Single-point coverage, as opposed to continuous coverage of multiple rooms, is available	The WAC+Fit AP architecture supports continuous AP networking with functions such as automatic calibration and roaming

1.1.3 Phase 3: All-Wireless Office Era — All-Wireless Office, Wireless-Centric

Currently, enterprise WLANs have entered the third phase. In enterprise office environments, wireless networks are used in preference to wired networks, and each office area is covered entirely by a Wi-Fi network. Furthermore, office areas do not include a wired network port, making the office environment more open and intelligent. In the future, high-bandwidth services, including enterprise cloud desktop office, telepresence conference, and 4K video, will be migrated from wired to wireless networks. Likewise, new technologies such as virtual reality (VR)/augmented reality (AR), virtual assistant, and automatic factory will be directly deployed on wireless networks. These new application scenarios pose higher requirements for enterprise WLAN design and planning.

The year 2018 marked the release of the next-generation Wi-Fi standard, referred to as Wi-Fi 6 and 802.11ax by the Wi-Fi Alliance and IEEE, respectively. This represents another major milestone in the development of Wi-Fi. In that regard, the core value of Wi-Fi 6 is further improvements in capacity, leading indoor wireless communications into the 10-gigabit era. The concurrent performance of multiple users has improved fourfold, ensuring excellent service capabilities in high-density access and heavy-load scenarios.

Enterprise WLANs are also evolving toward WLAN-based positioning service functions and IoT connection functions. For example, high-precision indoor navigation and positioning are supported by integrating Wi-Fi and Bluetooth positioning technologies. In addition, short-range IoT technologies, such as radio frequency identification (RFID) and ZigBee, are used to support IoT applications, including electronic shelf labels and smart bands.

1.2 CHALLENGES FACED BY ENTERPRISE WLANS

Driven by the rapid iteration of standards, and accompanied by an explosive growth in demand scenarios, enterprise WLANs have evolved from wired networks to all-wireless office networks. Requirements for services carried on networks go beyond just Internet access. Enterprise WLANs have become the infrastructure that supports digital transformation across various industries while improving production and work efficiency. As a result, they face more difficult challenges, as illustrated in Figure 1.3.

1.2.1 WLAN Services Require Ultra-Large Bandwidth

In the future, all industries will undergo digital transformation. Taking enterprise offices as an example, mobility and cloudification are two important aspects of digital transformation. With a one-stop office platform that integrates instant messaging, email, video conferencing, and approval services, an increasing number of tasks can be processed on mobile devices anytime, anywhere. Mobile and cloud-based networks can help enterprises limit the effect of geographical distances and implement collaborative offices. Especially with the emergence of new technologies such as 4K telepresence conferences, people in different locations can participate in the same project, and team members can hold telepresence

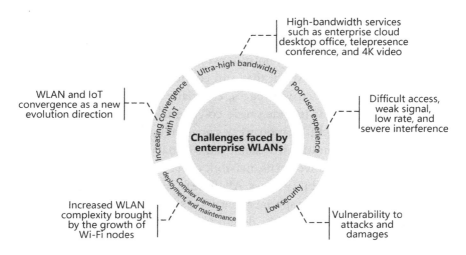

FIGURE 1.3   Challenges faced by enterprise WLANs.

conferences anytime, anywhere. With clear and smooth features, 4K video conference can generally create the same experience as a face-to-face conference. (Telepresence conferences have high requirements of a bandwidth at least 50 Mbps and a latency less than 50 ms.)

In addition, digital classroom has become a new trend in the education field. Online interactive teaching, HD video on demand (VOD), and academic video conferencing redefine the teaching and learning pattern. Students can use mobile terminals to learn anytime, anywhere, fully utilizing time that would otherwise be wasted. Furthermore, the emergence of VR technology brings with it immersive courses entailing high participation, thereby enhancing the teaching effect. Surgery demonstrations are typical examples where VR is used to teach. The 4K/VR surgery demonstration system is deployed to synchronize surgery procedures to each classroom in real time. Doctors in the operation room can communicate with consultation experts in real time through HD videoconferencing devices, which all require ultra-large bandwidth.

The bandwidth of a single user of traditional services does not exceed 10 Mbps. Therefore, a major challenge for enterprise WLANs is to meet the high-bandwidth requirements of concurrent services.

1.2.2 Wi-Fi Technology Bottlenecks Affect User Experience

In addition to bandwidth requirements, poor user experience is another factor affecting WLANs.

For example, in large exhibitions, organizers provide wireless Internet access for participants. However, exhibition participants find it difficult to connect to the provided Wi-Fi hotspot, which they can view in the search results on their devices. Even if they can connect to the Wi-Fi hotspot, it takes a long time to open web pages, especially when users are mobile. Possible reasons for this include severe co-channel interference caused by improper radio channel allocation, weak signal strength caused by improper channel power configurations, unstable radio signals caused by interference sources such as microwave ovens, network congestion caused by excessive users, and terminal capability issues. All these factors affect user experience. In addition, there are some special application scenarios in exhibitions. For example, some exhibition booths design interactive experiences using the VR technology, which poses high requirements on latency. Journalists may use wireless networks for live broadcast or text reporting. Furthermore, ensuring smoothness in these services is a challenge.

1.2.3 The Security of WLAN Services Is Questioned

WLANs use radio waves to transmit data, making wireless channels a major target for hackers and criminals who wish to attack and sabotage data transmission. Especially in the all-wireless office era where core services are carried by WLANs instead of wired networks, any loss or damage resulting from WLAN attacks will be devastating.

Take the financial industry as an example. According to the *China Banking Industry Service Report 2017* released by the China Banking Association, China's mobile banking business has grown rapidly, and the number of individual mobile banking customers reached 1502 million by 2017. Using wireless networks, people can perform operations such as query, transfer, and financial management on their accounts anytime, anywhere on mobile terminals. However, customers take risks when using mobile banking. Once wireless networks are attacked and data is stolen, direct economic losses will occur. In particular, Key Installation Attack (KRACK), which was widely reported on by the media in the second half of 2017, raised users' wireless security concerns.

1.2.4 The Growth of Network Devices Leads to More Complex Planning, Deployment, and Maintenance

A traditional wired network serves only users in a fixed area. The number of terminals in the area and network ports to be configured can be estimated. However, the situation becomes extremely complex when planning a wireless network. For example, a conference room with 50 people requires real-time video conferencing as well as full Wi-Fi coverage. In the traditional network solution, network administrators need to accurately estimate the service model of the coverage area, plan the Wi-Fi network, and configure radio frequency parameters to avoid co-channel interference between APs and ensure the coverage and quality of radio signals. To meet the requirements of real-time video conferencing on the bandwidth, delay, and packet loss rate, complex quality of service (QoS) parameter configurations are required. Finally, network administrators need to manually configure commands on each device separately. Therefore, if a configuration error occurs, network administrators need to check the commands accordingly, which is both time- and labor-consuming. In addition, once a wireless device is faulty, user access in an area is affected. Therefore, the working status of APs and switches needs to be monitored in real time. However, if there are a large number of APs, the manual monitoring workload is heavy.

1.2.5 Challenges Regarding WLAN and IoT Convergence

With the rapid growth of WLANs, IoT technologies are also developing quickly and being widely used. In the era with all things connected, enterprise IoT is ubiquitous. In enterprise office scenarios, IoT can be used for enterprise asset management to implement functions such as asset tracking and automatic counting. In schools, IoT can be used for student health management, and also for implementing automatic student attendance and student sign detection. In hospitals, IoT can be used for infusion management, medication management, and real-time monitoring of vital signs. In factories, IoT can be used for real-time interconnection of production resources in workshops to implement refined management and control of production resources.

WLAN and IoT technologies are developing rapidly, and the industry is exploring the possibility of integrating them. On the one hand, independent IoT deployment requires heavy investments, and it is complex to manage and maintain WLAN and IoT networks separately. On the other hand, IoT and WLAN have numerous similarities, including physical layer protocols, frequency bands, deployment, and networking modes. Therefore, IoT and WLAN are evolving from coexistence to convergence and ultimately normalization.

1.3 NEXT-GENERATION ENTERPRISE WLAN SOLUTIONS

To address the preceding challenges, the next-generation enterprise WLAN solutions have been launched, as illustrated in Figure 1.4.

1.3.1 Support for 802.11ax to Meet Ultra-High Bandwidth Requirements

With the popularization of new services such as 4K/8K HD video and AR/VR, high bandwidth and low latency are required for WLANs. To meet the requirements for ultra-high bandwidth, the 802.11 standard is dedicated to making breakthroughs in transmission rates, from initially only 11 Mbps in 802.11b, to more than 1 Gbps in 802.11ac, and 10 Gbps in the more recent 802.11ax (Wi-Fi 6). In addition, various multiuser technologies have been introduced to improve the actual throughput of users.

1.3.2 Innovative Wireless Air Interface Technology, Providing Excellent User Experience

The following key technologies need to be improved to provide optimal user experience:

1. Wireless coverage

   To achieve optimal network coverage, you must plan proper channels and power for a WLAN. The radio frequency (RF) calibration function can automatically deploy channels and adjust the transmit power, while also coping with diversified network environments,

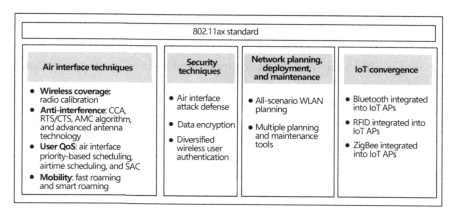

FIGURE 1.4   Next-generation enterprise WLAN solutions.

avoiding interference, and filling coverage holes. As a result, optimal wireless coverage can be achieved. In addition, RF calibration can automatically detect uncertain factors on a network and optimize wireless network parameter settings, adapting to network changes.

2. Anti-interference

One of the main WLAN features is that data is transmitted in open spaces, which include a large amount of interference. The next-generation enterprise WLAN solutions provide a series of technical measures to address interference. First, you can use the spectrum analysis function to identify non-Wi-Fi interference. Second, for Wi-Fi interference, you can reduce the probability of conflicts between wanted and interference signals by using the clear channel assessment (CCA) and request to send (RTS)/clear to send (CTS) mechanism. Finally, an adaptive modulation and coding (AMC) algorithm can reduce performance deterioration caused by inevitable interference.

In addition to using software algorithms to improve anti-interference capabilities, advanced antenna technologies also enable APs to avoid interference. On the one hand, high-density antennas can effectively control the coverage of APs while reducing interference between neighboring APs. On the other hand, smart antennas can ensure radio signal quality by using the beamforming technology.

3. QoS

Diversified user services require optimal user experience from differentiated WLAN services. In that regard, the next-generation enterprise WLAN solutions provide multiple QoS technologies for various services. Specifically, air interface priority-based scheduling helps preferentially schedule high-priority services. Airtime scheduling enables multiple users to fairly share radio resources and transmit more data. Smart application control (SAC) can further identify service types, especially high-priority services such as voice and video, and enable more refined control to be performed on those services.

4. Mobile experience

Unlike users in wired networks, users in WLANs do not have fixed locations. Therefore, it is important to ensure user experience during roaming. To that end, the fast roaming function ensures that services are not interrupted and reauthentication is not required when users move. This function also provides services for STAs that cannot roam smoothly, improving user experience during roaming.

1.3.3 Integrated Security Technologies to Enable Comprehensive Wireless Protection

With the evolution of 802.11 standards, WLAN security protection measures are continuously being upgraded.

Attaching high importance to network security, the next-generation enterprise WLAN solutions use diversified user authentication mechanisms and establish secure associations to ensure the legitimacy of all parties involved in communication. Specifically, wireless data is encrypted to ensure the security of wireless data links. Attacks from unauthorized terminals and malicious users, as well as intrusions to a WLAN, can be detected. Furthermore, wireless attack countermeasures can be used to prevent unauthorized wireless devices from accessing an enterprise network and defend against attacks on the network system.

1.3.4 Scenario-Based Network Planning, Deployment, and Maintenance

The next-generation enterprise WLAN solutions provide multiple networking schemes to meet requirements of enterprises and organizations. In addition, due to their distinctive features, WLANs require networking schemes as well as network planning. Network planning solutions need to extend common wireless network planning methods and be designed based on service characteristics, including service types, service usage frequency, user distribution, terminal types, key services, and key areas. Network planning is becoming more difficult due to the expansion of network scale. Professional tools can be introduced in the network planning, deployment, acceptance, and maintenance phases to implement one-click intelligent planning and deployment, simplifying the complex design process and shortening the project delivery time. When an exception occurs on the network or in the external environment, the system can automatically notify engineers of the exception, which significantly reduces maintenance costs.

1.3.5 WLAN and IoT Convergence Solution

The next-generation enterprise WLAN solutions also include the WLAN and IoT integration solution. For example, an IoT AP can use the Bluetooth positioning technology to support high-precision indoor navigation and positioning. Furthermore, an IoT AP can use short-range IoT technologies, including RFID and ZigBee, to support IoT applications, such as electronic shelf labels and smart bands.

CHAPTER 2

WLAN Technology Basics

S IMILAR TO OTHER WIRELESS technologies (such as radio broadcast-ing), wireless local area network (WLAN) enables the transmission of information by radio waves over the air. This chapter opens with the basic concepts of wireless communication and key WLAN technologies before moving on to the evolution and version comparison of IEEE 802.11 stan-dards, and finishes with a discussion of the next-generation 802.11ax in terms of key technologies at the physical and MAC layers in addition to basic principles.

2.1 CONCEPT OF WIRELESS COMMUNICATION

Wireless communication involves the transmission of information over the air by radio waves, which radiate and propagate through space.

2.1.1 Radio Waves

Radio waves are a type of electromagnetic wave capable of transmitting energy and momentum over the air through the use of electric and mag-netic fields whose oscillations are in the same phase but both perpendicular to each other and to the direction of propagation. Electromagnetic waves travel at the speed of light. Figure 2.1 illustrates electromagnetic waves.

Electromagnetic waves are continuously emitted from the everyday objects around us and propagate outward in all directions where they are reflected, refracted, or scattered. In wireless communication, multiple identical electromagnetic waves emitted from a transmitter itself (known as multipath interference) due to direct radiation (LOS propagation) and

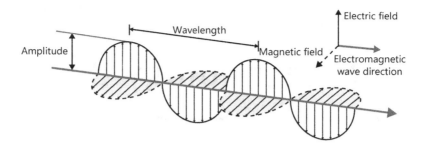

FIGURE 2.1   Electromagnetic waves.

reflection, in addition to the electromagnetic waves originating from other sources, result in interference on the receiver end, as illustrated in Figure 2.2. As such, wireless communication is far more complex than traditional wired communication.

Frequency is an important parameter of electromagnetic waves and its distribution is referred to as "spectrum." In order of decreasing frequency, electromagnetic waves can be classified into gamma rays, X-rays, ultraviolet rays, visible lights, infrared rays, microwaves, and radio waves, as illustrated in Figure 2.3. Higher-frequency electromagnetic waves are more energetic, exhibit stronger direct radiation capabilities, but feature faster energy attenuation during transmission and shorter transmission distances.

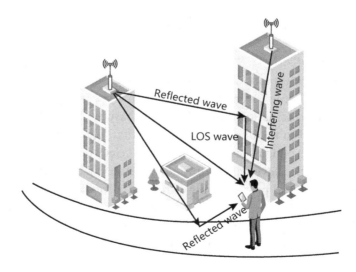

FIGURE 2.2   Multipath interference and external interference.

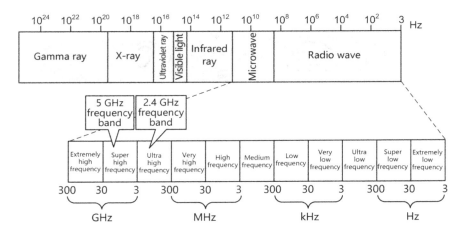

FIGURE 2.3   Electromagnetic wave spectrum.

WLANs utilize radio waves. A radio wave is generated by an oscillation circuit's alternating current and can be transmitted and received over an antenna. It is also referred to as radio, electric wave, or radio frequency (RF).

The frequency range of a radio wave is known as the frequency band. WLANs operate in the 2.4 GHz (2.4–2.4835 GHz) or 5 GHz (5.15–5.35 GHz or 5.725–5.85 GHz) frequency band. Designed for Industrial, Scientific, and Medical (ISM) purposes, these frequency bands can be used without obtaining license or paying fees as long as the transmit power requirement (generally less than 1 W) is met and no interference is caused to other frequency bands. While such cost-free resources reduce WLAN deployment costs, co-channel interference can arise when multiple wireless communications technologies operate within the same frequency band. Available ISM frequency bands vary depending on country and region, and WLANs are required to use frequency bands in compliance with local laws and regulations.

2.1.2 Wireless Communications System

Before we explore how information is transmitted through radio waves, let's take a look at an example of sound transmission, as this will help understand the composition of a wireless communications system. When A says "Hello" to B, A is actually sending an instruction to the mouth to generate sound, which then converts the information into a group of sound waves with a specific frequency and amplitude. The sound waves vibrate and propagate through space until they reach B's ears and are converted into identifiable information. See Figure 2.4.

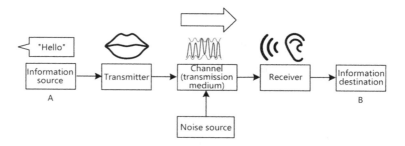

FIGURE 2.4 Communications system.

The sound transmission process in our example is a type of communications system. A is the information source, and "Hello" is the information to be transmitted. Here, A's mouth is the transmitter used to convert the information into sound waves by means of modulation. The waves are then transmitted over the air, which functions as the channel (or transmission medium). B's ears serve as the sound wave receiver, and B as the information destination converts sound waves into identifiable information by means of demodulation. During the sound wave transmission, speech from other people may result in noise or interference, possibly leading to B failing to clearly hear A's message.

In a wireless communications system, information to be transmitted can include images, text, or sound. A transmitter applies source coding to first convert the information into digital signals that allow for circuit calculation and processing, and then into radio waves by means of channel coding and modulation, as illustrated in Figure 2.5. Transmitters and receivers are connected through interfaces and channels. Unlike the visible interfaces involving cable connections used for wired communication, wireless interfaces are invisible and connected over the air. As such, a wireless interface is referred to as an air interface.

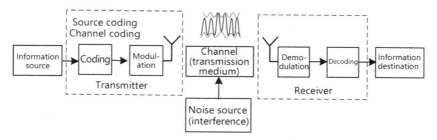

FIGURE 2.5 Wireless communications system.

2.1.2.1 Source Coding

Source coding is the process of converting raw information into digital signals by using a coding scheme. Source coding can minimize redundant raw information by compressing the information to the maximum extent without causing distortion. Different types of information require different coding schemes (e.g., H.264 for videos).

2.1.2.2 Channel Coding

Channel coding is a technology used for detecting and correcting information errors in order to improve channel transmission reliability. As we have seen in our example, wireless transmission is prone to noise interference, and this can lead to errors in the information delivered to the receiver. Channel coding is utilized to restore information to the maximum extent at the receiver end, thereby reducing the bit error rate. Binary convolutional coding (BCC) and low-density parity-check (LDPC) code can be used for WLAN channel coding.

Channel coding adds redundant data to the raw information, increasing the information length. The ratio of the number of precoding bits (raw information) to the number of postcoding bits is referred to as the coding efficiency. Channel coding decreases the transmission rate but increases the transmission success rate of valid information. As such, selecting a proper coding scheme for communication protocols is crucial to achieving an optimal tradeoff between performance and accuracy.

2.1.2.3 Modulation

Digital signals in circuits are instantaneous changes between high and low levels. Only after being superimposed on high-frequency signals generated by high-frequency oscillation circuits, the digital signals can be converted into radio waves over antennas and then transmitted. This process of superimposition is known as modulation. Having no information itself, a high-frequency signal instead carries information and is therefore called a carrier.

The modulation process encompasses both symbol mapping and carrier modulation.

Symbol mapping involves mapping bits of digital signals to symbols (also referred to as code elements or information elements) by modulation, with each symbol representing one or many bits. For example, one symbol carries one bit, two bits, and four bits with binary phase shift keying

(BPSK), quadrature phase shift keying (QPSK), and 16 quadrature amplitude modulation (16QAM), respectively. A larger quantity of bits carried in a single symbol indicates a higher data transmission rate.

Carrier modulation refers to the process of superimposing symbols and carriers, ensuring that the carriers are carrying the information to be transmitted. Multicarrier modulation can be used to further improve the transmission rate and works by taking advantage of the orthogonal characteristics of waves to split signals into several groups, with each modulated over separate carriers, and then combine the modulated signals for transmission over antennas, achieving concurrent transmission of multiple groups of signals. Multicarrier modulation can effectively utilize spectrum resources and reduce multipath interference. Orthogonal frequency division multiplexing (OFDM) is the multicarrier modulation technology used in WLANs.

2.1.2.4 Channel

A channel transmits information, and a radio channel is a radio wave in space. Given that radio waves are ubiquitous, random use of spectrum resources will result in endless interference issues. Therefore, in addition to defining usable frequency bands, wireless communication protocols must also accurately divide frequency ranges, with each frequency range known as a channel.

For example, the WLAN 2.4 GHz frequency band is divided into 14 channels with overlapping or nonoverlapping relationships, each with a bandwidth of 22 MHz, as illustrated in Figure 2.6.

- Overlapping channels, such as channel 1 and channel 2, interfere with each other.

- Nonoverlapping channels, such as channel 1 and channel 6, can coexist in the same space without interfering with each other.

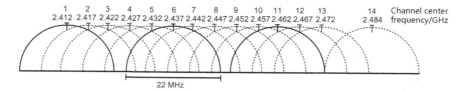

FIGURE 2.6   Channel division in the 2.4 GHz frequency band.

2.1.3 Channel Rate and Bandwidth

Wireless communication requires both a high channel transmission rate and a low error rate. The channel transmission rate, in unit of bps, is referred to as the data or bit rate, while the maximum rate that can be achieved on a channel is known as the channel capacity or throughput.

Bandwidth, also referred to as frequency or spectrum width, is a contiguous frequency range representing the amount of spectrum resources available (displayed in units of Hz). In some cases, bandwidth can also indicate the channel capacity. This is in accordance with the Nyquist theorem: in an ideal situation where noise can be ignored, if the channel bandwidth is B (unit: Hz), the symbol rate will be B multiplied by 2 (unit: Baud) and then the channel capacity can be obtained based on the number of bits mapped to symbols. For example, if the frequency range of one channel is 5170–5190 MHz and one symbol carries six bits, the bandwidth is 20 MHz and the maximum theoretical rate is 240 Mbps. As a result, a bandwidth can indicate a rate and a spectrum segment width.

While the Nyquist theorem describes a perfect channel without noise, in reality, noise is omnipresent. The Shannon theorem tells us that the actual channel capacity is susceptible to noise, the impact of which may be represented by the signal-to-noise ratio (SNR), that is, the ratio of signal power to noise power. For a given bandwidth, larger noise indicates less channel capacity. However, according to the Shannon theorem, when the bandwidth tends to infinity, channel capacity does not increase infinitely.

2.2 WLAN KEY TECHNOLOGIES

As a wireless communication technology, the following WLAN improvements are crucial to delivering higher transmission rates:

- Upgraded modulation scheme

- Enhanced number of carriers by using OFDM

- Channel bonding to expand channel bandwidth

- Increased number of spatial streams by using multiple-input multiple-output (MIMO)

This section describes the technologies used to improve wireless rates, including modulation schemes, OFDM, channel bonding, and MIMO.

To ensure consistent Internet service experience in the event of multiuser access, WLAN also introduces multiple access techniques to distinguish individual users, such as multiuser MIMO (MU-MIMO) and orthogonal frequency division multiple access (OFDMA).

Wireless communication differs from wired communication in a number of key areas. In wired communication, the signal collision between transmitters can be detected using high and low levels on cables. Such detection is not possible in wireless communication. Instead, 802.11 standards have designed a simple distributed access protocol at the Media Access Control (MAC) layer: carrier sense multiple access with collision avoidance (CSMA/CA). This section describes the technology in detail.

2.2.1 Modulation Technology

Based on the three parameters of electromagnetic waves (amplitude, frequency, and phase), digital signal modulation technologies can be classified into amplitude shift keying (ASK), frequency shift keying (FSK), and phase shift keying (PSK), as illustrated in Figure 2.7. Quadrature amplitude modulation (QAM) is another modulation scheme that combines ASK and PSK and delivers the highest modulation efficiency. Modulation enables the superimposition of digital signals that leads to slight changes in radio waves.

- ASK: Variations in the amplitude of a carrier represent 0 and 1. While this scheme is easy to implement, it offers poor anti-interference performance.

- FSK: Variations in the frequency of a carrier represent 0 and 1. This scheme, used for low-speed data transmission, is easy to implement and delivers strong anti-interference capabilities.

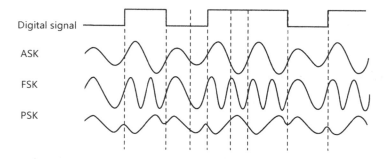

FIGURE 2.7 Modulation schemes.

- PSK: 0 and 1 represent variations in the phase of a carrier. This scheme is also referred to as *M*PSK, with *M* representing the number of symbols, such as BPSK (2PSK), QPSK (4PSK), 16PSK, and 64PSK. Of these, BPSK and QPSK are most common. For example, the simplest BPSK uses phases 0° and 180° to indicate 0 and 1, respectively, meaning two symbols transmit one bit. QPSK uses the four phases 0°, 90°, 180°, and 270° to represent four symbols 00, 01, 10, and 11, respectively. Here, each symbol carries two bits, twice that of BPSK. See Figure 2.8.

- QAM: Two orthogonal carriers are used for amplitude modulation in order to transmit more information on a symbol. The format of *N*QAM can be used, with *N* representing the number of symbols, such as 16QAM, 64QAM, 256QAM, and 1024QAM. A larger value for *N* indicates a larger data transmission rate and a higher bit error rate.

WLAN signals can be modulated in BPSK, QPSK, or QAM mode. The following examples mainly depict QAM.

Comparing BPSK and QPSK confirms that the number of bits carried on a symbol increases with the number of phases. However, it is not feasible to continuously increase the number of phases without limits. Once a certain point is reached, adjacent phases begin to offer extremely small differences, which reduces the anti-interference capabilities of modulated signals. This is where QAM comes in.

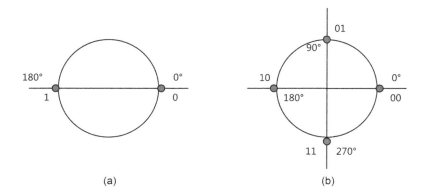

(a)   (b)

FIGURE 2.8   (a) BPSK and (b) QPSK modulation.

16QAM, for example, adjusts the amplitude and phase of a carrier to generate 16 waveforms (representing 0000, 0001, and so on). As such, one symbol carries four bits, as illustrated in Figure 2.9.

16QAM is implemented as follows: The data $(S_3S_2S_1S_0)$ obtained from channel coding is mapped to complex modulated symbols on a constellation diagram. Each symbol's I and Q components (corresponding to the real and imaginary parts of the complex plane, which relate to the horizontal and vertical directions, respectively) are modulated by amplitude onto two carriers with orthogonality in the time domain ($\cos\omega_0 t$ and $\sin\omega_0 t$) and finally superposed to generate modulated signals, as illustrated in Figure 2.10.

A constellation diagram illustrates the mapping among the input data, in-phase/quadrature (I/Q) data, and carrier phase. Figure 2.11 illustrates the 16QAM constellation diagram.

A constellation diagram uses polar coordinates, with each point represented by an angle and distance from the point to the origin, which correspond to the phase and the amplitude by which QAM is implemented, respectively.

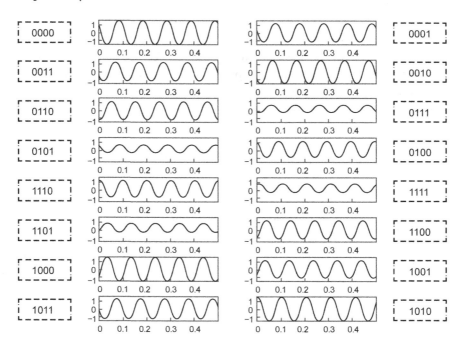

FIGURE 2.9   16QAM.

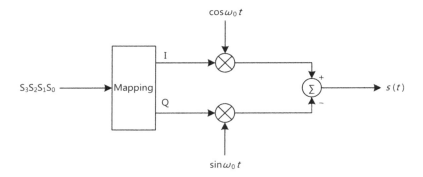

FIGURE 2.10   QAM process.

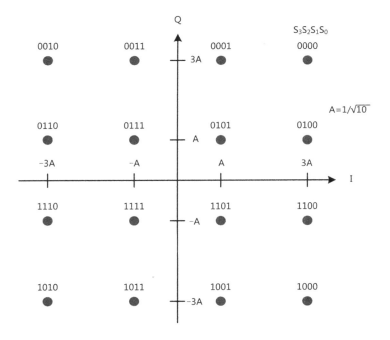

FIGURE 2.11   16QAM constellation diagram.

Table 2.1 lists the transmission rates that vary with coding rates and modulation schemes, as stipulated in 802.11a.

Table 2.1 confirms that QAM with more points in the constellation diagram can be used to improve transmission rates. QAM featuring 1024 constellation points (1024QAM) is currently available.

TABLE 2.1   Various Rates in Modulation Schemes in 802.11a

Modulation Scheme	Coding Rate	Rate (Mbps^{-1})
BPSK	1/2	6
BPSK	3/4	9
QPSK	1/2	12
QPSK	3/4	18
16QAM	1/2	24
16QAM	3/4	36
64QAM	2/3	48
64QAM	3/4	54

TABLE 2.2   MCS Indexes for 802.11ac

MCS Index	Modulation Scheme	Coding Rate
0	BPSK	1/2
1	QPSK	1/2
2	QPSK	3/4
3	16QAM	1/2
4	16QAM	3/4
5	64QAM	2/3
6	64QAM	3/4
7	64QAM	5/6
8	256QAM	3/4
9	256QAM	5/6

However, an increase in constellation points does not always equate to improved performance. As the interpoint distance decreases with the number of points, higher requirements are placed on the receive signal quality to prevent demodulation errors.

The modulation and coding scheme (MCS) is also introduced in 802.11 standards. Using 802.11ac as an example, 10 MCS indexes are available, each representing a group of MCSs, as listed in Table 2.2. For any given MCS index, the rate can vary with the channel bandwidth, number of spatial streams, and guard interval (GI).

2.2.2 OFDM

2.2.2.1 Principles

OFDM is a multicarrier modulation technology. By dividing a channel into multiple orthogonal subchannels, it converts high-speed serial data signals into low-speed parallel data signals and modulates them onto the subchannels for transmission. Carriers corresponding to orthogonal subchannels are usually referred to as subcarriers.

High-speed serial data is converted to low-speed parallel data, as the former can cause inter-symbol interference in wireless transmission.

Dividing a channel into orthogonal subchannels improves spectrum utilization, as this technique eliminates mutual interference and allows subcarriers to be as close as possible, or even superimposed, as illustrated in Figure 2.12.

Figure 2.12 illustrates a channel divided into three orthogonal subcarriers. The individual peak of each subcarrier is used for data coding and line up with zero amplitudes of the other two subcarriers. Given that the frequency at which OFDM signals need to be extracted for demodulation is exactly the maximum amplitude of each subcarrier, extracting signals from these overlapping subcarriers does not result in interference.

In an actual WLAN, OFDM subcarriers are divided in compliance with specific rules. Using 802.11a as an example, OFDM divides a 20 MHz channel in the 5 GHz frequency band into 64 subcarriers, each with a frequency width of 312.5 kHz. Figure 2.13 illustrates OFDM symbols in a PLCP Protocol Data Unit (PPDU).

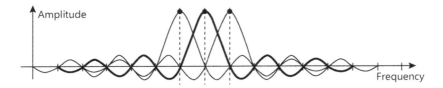

FIGURE 2.12   Example of orthogonal subcarriers.

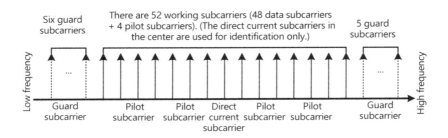

FIGURE 2.13   Subcarrier division in 802.11a.

- Guard subcarrier: used as a GI to reduce interference between adjacent channels. A guard subcarrier carries no data. There are six guard subcarriers on the left (the low frequency side) and five guard subcarriers on the right.

- Pilot subcarrier: used to estimate channel parameters and carry specific training sequences in certain data demodulation. There are four pilot subcarriers.

- Direct current subcarrier: located at the spectrum center and usually idle (used for identification only).

- Data subcarrier: used to transmit data. 802.11a and 802.11g stipulate 48 data subcarriers, while 802.11n and 802.11ac stipulate 52.

Generally, working subcarriers include pilot subcarriers and data subcarriers. The number of working subcarriers in 802.11a is 52 (4 + 48).

OFDM uses inverse fast Fourier transformation (IFFT) for modulation and fast Fourier transformation (FFT) for demodulation, as illustrated in Figure 2.14.

The transmitter processes a batch of parallel signals each time, uses IFFT to modulate these raw signals to corresponding subcarriers, and then superposes the signals to generate OFDM symbols in the time domain. The receiver performs an inverse operation (specifically FFT) on the received OFDM signals to correctly demodulate the data.

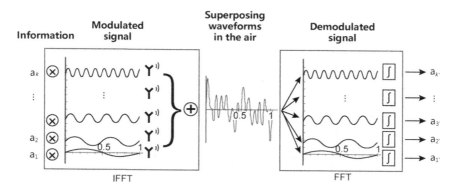

FIGURE 2.14 OFDM modulation and demodulation.

2.2.2.2 Advantages and Disadvantages
OFDM offers the following advantages:

1. High transmission rate

    In line with 802.11 standards, the maximum transmission rate delivered by traditional physical-layer technologies frequency hopping spread spectrum (FHSS) and direct sequence spread spectrum (DSSS) are 2 and 11 Mbps, respectively. With OFDM, the maximum transmission rate can be increased significantly, for example, reaching up to 54 Mbps when using 802.11a and 64QAM.

2. High spectral efficiency

    The orthogonal nature of subcarriers eliminates mutual interference and allows adjacent subcarriers to be close enough, ensuring higher spectral efficiency.

3. Anti-multipath interference

    OFDM divides a channel into multiple subcarriers. Intersymbol interference (ISI) due to multipath transmission is eliminated as long as the frequency width of each subcarrier is less than the related channel bandwidth.

4. Anti-fading capability

    OFDM delivers strong anti-fading capabilities due to the joint coding of subcarriers, leading to improved system performance. As channel-specific frequency diversity is already utilized by OFDM, it is not necessary to add a time-domain equalizer if severe fading is not an issue.

5. Anti-narrowband interference capability

    OFDM can resist narrowband interference to some extent because this type of interference only affects a small part of subcarriers.

However, OFDM also has the following disadvantages:

1. Susceptible to frequency offset

    OFDM has strict requirements on orthogonality between subchannels. Any spectrum offset of radio signals during transmission or frequency offset between the transmitter and the local oscillator of the receiver can devastate the orthogonality between subcarriers in the OFDM system, leading to interference between subchannels.

2. Relatively high peak to average power ratio (PAPR)

The OFDM system outputs the superimposition of multi-subcarrier signals. If these signals are in the same phase, the instantaneous power of the superimposed signals will be far higher than the average power of the signals. This results in a relatively large PAPR that is likely to trigger signal distortion, destroying inter-subcarrier orthogonality and generating interference.

2.2.2.3 Applications

OFDM is now a widely applied core technology in 3G and 4G mobile communications. 802.11a, 802.11g, 802.11n, and 802.11ac all support OFDM to improve transmission rates.

2.2.3 Channel Bonding

Broadening the spectrum will improve wireless transmission rates, similar to how widening a road will increase transportation capacity. See Figure 2.15.

802.11a and 802.11g use channels with 20 MHz bandwidth. 802.11n allows two adjacent 20 MHz channels to be bound into a 40 MHz channel, effectively improving transmission rates.

Channel bonding can more than double the transmission rates for spatial streams. A small portion of the bandwidth on a 20 MHz channel is reserved on both sides as GIs in order to reduce interference between adjacent channels. With channel bonding, the reserved bandwidth can also be used for communication. As such, with the 40 MHz bandwidth derived from channel bonding, the number of available subcarriers increases from 104 (52 × 2) to 108, resulting in a 208% increase in transmission efficiency, compared to that offered by the 20 MHz bandwidth.

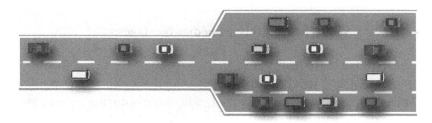

FIGURE 2.15 Relationship between vehicle transportation capacity and road width.

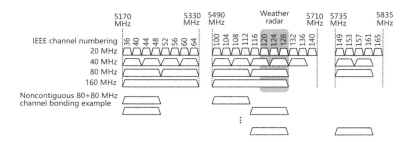

FIGURE 2.16   Channel bonding.

802.11ac also introduces channels with 80 MHz, 80+80 MHz (non-contiguous, nonoverlapping), and 160 MHz bandwidths, as illustrated in Figure 2.16.

An 80 MHz channel results from bonding two contiguous 40 MHz channels, and a 160 MHz channel from bonding two contiguous 80 MHz channels. Given the scarcity of contiguous 80 MHz channels, two non-contiguous 80 MHz channels can also be bonded to obtain a bandwidth of 160 MHz (80+80 MHz).

In the case of bound channels, 802.11 standards stipulate that one will function as the primary channel and the other as the secondary channel. Management packets, such as Beacon frames, must be sent on the primary channel rather than the secondary channel. In a 160 MHz channel, a 20 MHz channel must be selected as the primary channel. The 80 MHz channel containing this 20 MHz channel is the primary 80 MHz channel and the one without this 20 MHz channel is the secondary 80 MHz channel. In the primary 80 MHz channel, the primary 40 MHz channel contains this 20 MHz channel while the secondary 40 MHz channel does not. The remaining portion in the primary 40 MHz channel is called the secondary 20 MHz channel. See Figure 2.17.

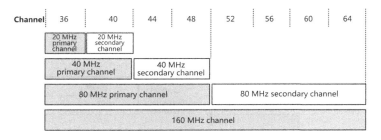

FIGURE 2.17   Primary/Secondary channel division.

802.11ac supports channel bandwidths ranging from 20 to 160 MHz, and this flexibility complicates channel management. Reducing inter-channel interference and maximizing channel utilization on a network where devices use channels with different bandwidths is a significant challenge.

802.11ac defines an enhanced request to send (RTS)/clear to send (CTS) mechanism to coordinate which channels are available, and when.

As illustrated in Figure 2.18, 802.11n utilizes a static channel management mechanism, where a busy subchannel will cause the entire bandwidth to become unavailable. 802.11ac uses a dynamic spectrum management mechanism. The transmitter sends RTS frames containing frequency bandwidth information on each 20 MHz channel, and the receiver determines whether the channels are busy or idle before responding with CTS frames only on available channels, which must encompass the primary channel. These CTS frames contain channel information. If the transmitter determines that some channels are extremely busy, it halves the transmit bandwidth as a result of dynamic management.

Dynamic spectrum management improves channel utilization and reduces interference between channels, as illustrated in Figure 2.19. With this mechanism, two APs can work on the same channel concurrently.

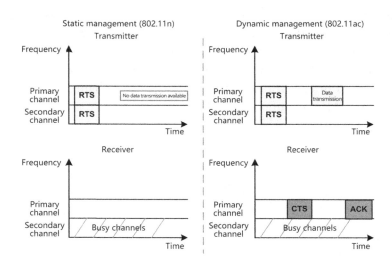

FIGURE 2.18 Spectrum management.

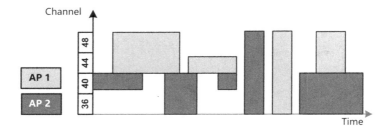

FIGURE 2.19   Two APs working on the same channel.

2.2.4 MIMO

2.2.4.1 Basic Concepts

Before using MIMO, we first need to understand single-input single-output (SISO). As illustrated in Figure 2.20, SISO establishes a unique path between the transmit and receive antennas, and transmits one signal between the two antennas. In the wireless system, each signal is defined as one spatial stream.

However, data transmission that uses SISO can be unreliable and rate limited.

To address this issue, in single-input multiple-output (SIMO), antennas are added on the receiver (terminal) side to allow two or more signals to be received concurrently, as illustrated in Figure 2.21.

In SIMO, two signals are sent from one transmit antenna carrying the same data. If one signal is partially lost during transmission, the receiver

FIGURE 2.20   SISO.

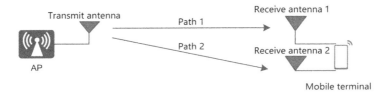

FIGURE 2.21   SIMO.

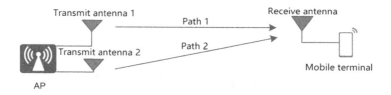

FIGURE 2.22   MISO.

can obtain the rest of the data using the other signal. In contrast to SISO that uses only one transmission path, SIMO ensures better transmission reliability with the transmission capacity unchanged. SIMO is also referred to as receive diversity.

Figure 2.22 illustrates two transmit antennas and one receive antenna.

In Figure 2.22, because there is only one receive antenna, the signals sent from the two transmit antennas are combined into one signal on the receiver. This mode, also referred to as multiple-input single-output (MISO) or transmit diversity, delivers the same effect as SIMO.

As the above analysis of SIMO and MISO proves that transmission capacity relies heavily on the number of transmit and receive antennas, it would be ideal to employ two antennas on both the transmitter and receiver sides to double the transmission rates by transmitting and receiving two signals separately.

As illustrated in Figure 2.23, MIMO is a transmission mode that uses multiple antennas on both the transmitter and receiver sides. MIMO enables the transmitting and receiving of multiple spatial streams (multiple signals) over multiple antennas concurrently, and it also differentiates between the different signals. Using technologies such as space diversity and spatial multiplexing, MIMO boosts system capacity, coverage scope, and SNR without increasing the occupied bandwidth.

The two key technologies used by MIMO are space diversity and spatial multiplexing.

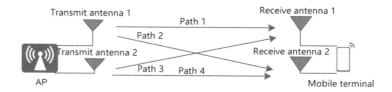

FIGURE 2.23   MIMO.

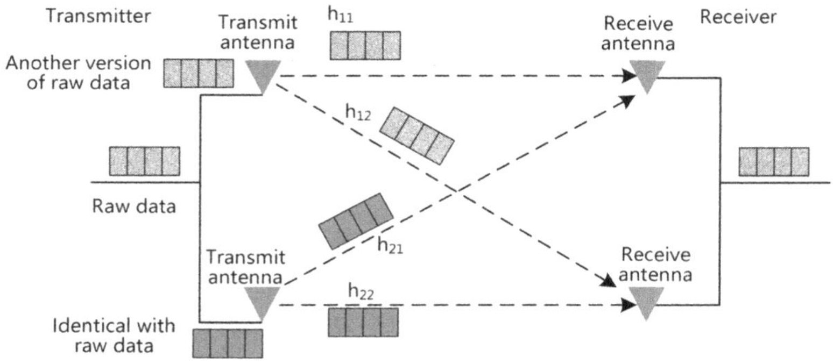

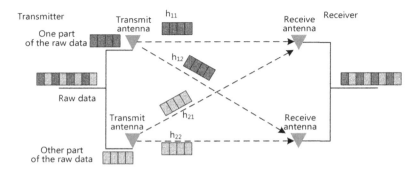

FIGURE 2.24 Space diversity.

Space diversity is a reliable transmission technique that is used to generate different versions of the same data stream and render them encoded and modulated before transmitting them on different antennas, as illustrated in Figure 2.24. This data stream can be either a raw data stream to be sent or a new data stream as a result of certain mathematical transformation on a raw data stream. The receiver uses a spatial equalizer to separate the received signals and performs demodulation and decoding to combine the signals of the same data stream and restore the original signals.

Spatial multiplexing divides the data to be transmitted into multiple data streams and renders them encoded and modulated before transmitting them over different antennas. This improves data transmission rates. Antennas are independent of each other. One antenna functions as an independent channel. The receiver uses a spatial equalizer to separate the received signals and performs demodulation and decoding to combine the data streams and restore the original signals, as illustrated in Figure 2.25.

FIGURE 2.25 Spatial multiplexing.

Multiplexing and diversity both involve space-time coding technology to convert one channel of data into multiple channels of data.

2.2.4.2 Relationship between Spatial Streams and the Number of Antennas

A MIMO system is generally written as $M \times N$ MIMO, with M and N indicating the number of antennas used for transmission and reception, respectively. It can also be written as $MTNR$, with T and R indicating transmission and reception, respectively. The number of spatial streams in MIMO is usually less than or equal to the number of antennas used for transmission or reception (the smaller of the number for transmission and that for reception prevails). For example, 4×4 MIMO (4T4R) enables the transmission of four or fewer spatial streams, and 3×2 MIMO (3T2R) enables the transmission of one or two spatial streams.

2.2.4.3 Beamforming

When the same signals are concurrently transmitted over multiple antennas in a wireless system, spatial holes may be generated due to transmission concurrency and multipath interference. For example, if signals, after being reflected by walls and devices, arrive at a location with two paths having the same attenuation but opposite phases, the two paths offset each other, leading to a spatial hole, as illustrated in Figure 2.26.

802.11n has proposed that beamforming technology be used to avoid spatial holes. Beamforming superimposes two beams by precompensating the phases of transmit antennas. Beamforming uses the weights derived

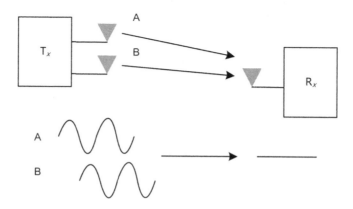

FIGURE 2.26 Spatial hole due to opposite phases.

based on the transmission environment or channel state information (CSI) to improve reception.

2.2.4.4 MU-MIMO

MIMO can be classified into single-user MIMO (SU-MIMO) and MU-MIMO by the number of users it allows to simultaneously transmit or receive multiple data streams.

SU-MIMO increases the data transmission rate of a single user by transmitting multiple parallel spatial streams over the same time-frequency resources to the user. In Figure 2.27, SU-MIMO is implemented between an AP with four antennas and a user with two antennas.

MU-MIMO increases the transmission rates of multiple users by transmitting multiple parallel spatial streams over the same time-frequency resources to multiple different users. In Figure 2.28, MU-MIMO is implemented between an AP with four antennas and four users each with one receive antenna.

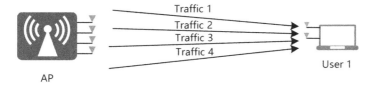

FIGURE 2.27   SU-MIMO.

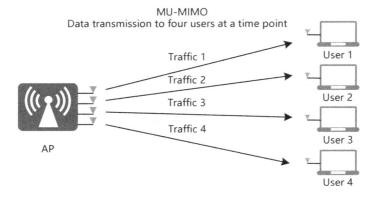

FIGURE 2.28   MU-MIMO.

Space division multiple access (SDMA) is essential to MU-MIMO. SDMA transmits multiuser data over the same timeslot and subcarrier but using different antennas to differentiate users spatially, thereby accommodating more users and increasing link capacity.

2.2.5 OFDMA

Before proceeding to OFDMA, we need to first understand multiple access. Multiple access is a technology used to distinguish users. For example, by assuming that an AP receives information from multiple users simultaneously, multiple access technology is used to distinguish these users in order for the AP to correctly respond to each user. SDMA for MU-MIMO is also a multiple access technology.

OFDMA is a combination of Frequency Division Multiple Access (FDMA) and OFDM.

FDMA differentiates users by frequency at a specific time point, as illustrated in Figure 2.29.

If we were to compare the presence of different frequencies to a classroom environment, FDMA would be used to enable different students (users) to learn in different classrooms at the same time.

OFDMA is an orthogonal frequency division multiple access technology that distinguishes different users by frequency. In FDMA, each frequency represents a user, with GIs required between frequencies to reduce interference, as illustrated in Figure 2.30. By contrast, OFDMA uses orthogonal frequencies whose orthogonality, as described in Section 2.2.2, eliminates mutual interference as well as the need for GIs. Such frequencies can even

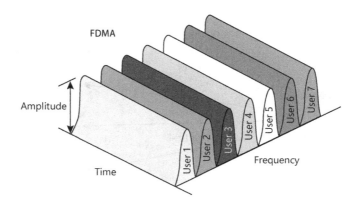

FIGURE 2.29   FDMA.

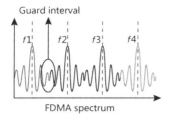

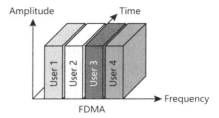

FIGURE 2.30   FDMA spectrum utilization.

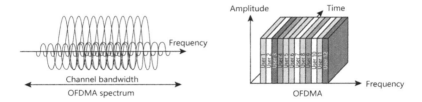

FIGURE 2.31   OFDMA spectrum utilization.

overlap with each other, as illustrated in Figure 2.31. Therefore, OFDMA offers significantly improved spectrum utilization when compared with FDMA.

What is the relationship between OFDMA and OFDM?

In Figure 2.32, in the presence of four users, the OFDM working mode is illustrated on the left, while the OFDMA working mode is illustrated on the right. The x-axis (t) represents the time domain, while the y-axis (f) represents the frequency domain (corresponding to different subcarriers).

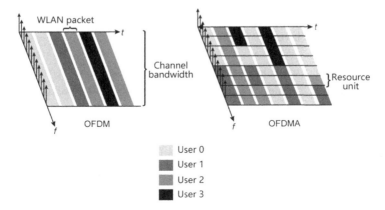

FIGURE 2.32   Comparison between OFDMA and OFDM.

In the OFDM working mode, the four users occupy channel resources separately in different timeslots. A user, in a specific timeslot, uses all subcarriers exclusively to transmit data packets (the WLAN packets in Figure 2.32).

By contrast, in the OFDMA working mode, channel resources are allocated to the four users by time-frequency resource unit (RU). The entire channel resources are divided into time-frequency RUs, where the data of each user is carried on their respective RUs. The OFDMA working mode allows more users to transmit data at a time point when the total time-frequency resources remain unchanged.

OFDMA has two advantages over OFDM:

- It allocates resources more appropriately. That is, when the channel quality of some users is poor, the transmission power can be allocated based on channel quality, thereby ensuring the proper allocation of channel time-frequency resources.

- It provides better quality of service (QoS). In OFDM, a single user occupies a whole channel, while another user with QoS-characterized data packets to be sent has to wait until the channel is released by the previous user, causing long delays. In OFDMA, however, a transmitter occupies only part of a channel, reducing the access delay of QoS nodes.

2.2.6 CSMA/CA

Even though a WLAN channel is shared by all STAs, data can only be transmitted by one STA in a period of time. Therefore, a channel allocation mechanism is required to coordinate when each STA transmits and receives data. 802.11 standards proposed the following two coordination modes at the MAC layer:

- Distributed coordination function (DCF): The CSMA/CA mechanism is used to enable each STA to contend for data transmission on a channel.

- Point coordination function (PCF): Centralized access control is used to enable STAs to transmit data frames when it is their turn (in a way similar to polling), thereby preventing conflicts.

DCF is mandatory and PCF is optional. DCF is most commonly used in the industry due to its CSMA/CA mechanism.

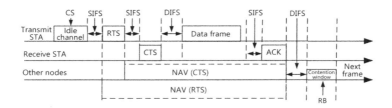

FIGURE 2.33   CSMA/CA mechanism.

With CSMA/CA, an STA listens to the channel status before transmitting data. If the channel is idle, indicating the absence of conflicts with other STAs, the STA is allowed to transmit data. However, if the channel is occupied, the STA waits before it transmits data. CSMA/CA includes carrier sense (CS), interframe space (IFS), and RB, as illustrated in Figure 2.33.

2.2.6.1 CS

The CS process involves monitoring a channel to determine whether the channel is busy or idle. In line with wireless channel characteristics, 802.11 standards proposed physical and virtual CS modes that run concurrently. If a wireless channel is determined to be busy by either mode, the channel state is busy.

Physical CS works at the physical layer and relies on the signal transmission medium and modulation mode. In this mode, after receiving a signal, an antenna detects the signal energy to estimate whether the channel is idle or busy. This mechanism is also referred to as clear channel assessment (CCA).

CCA uses two thresholds: signal detect (SD) and energy detect (ED). If we were to compare the CCA mechanism to a conversation among multiple people, the SD threshold is used to check for the presence of the speaker. Then, only after the current speaker finishes speaking, it will enable another person to speak. The ED threshold, however, detects whether the environment is too noisy, as nobody can be heard in a noisy environment, and ensures that participants only start speaking when their environment is quiet.

Virtual CS works at the MAC layer. In this mode, STAs are simply notified of the duration in which a channel is busy, rather than actively listening on a physical channel.

To reduce the extra consumption caused by data frame conflicts and hidden nodes, a transmit STA obtains the use rights of a channel through the RTS/CTS mechanism before transmitting data. (This process is explained in Section 3.5.3.) As illustrated in Figure 2.33, the transmit STA first sends an RTS frame carrying a channel occupation duration, also called the network allocation vector (NAV). Upon reception of the RTS frame, the surrounding STAs will not transmit data on the channel during the specified duration. In response to the RTS frame, the receive STA replies with a CTS frame that also carries an NAV. This means that the surrounding STAs that receive this CTS frame will also not transmit data during this duration. The STAs surrounding the transmit and receive STAs use their latest NAVs as a countdown timer. That is, when the countdown reaches 0, the channel is considered idle.

2.2.6.2 IFS

To avoid collisions, 802.11 standards require that all STAs wait before transmitting frames at the MAC layer. During this period, also called the IFS, the STAs still listen to the channel state. The IFS length depends on the type of frame to be sent. Higher-priority frames have a shorter IFS, meaning they can be sent earlier and occupy the channel. With the channel state busy, the transmission of low-priority frames is delayed, reducing the chance of collision.

Common IFS solutions are described as follows:

Short interframe space (SIFS) is used to separate frames of the same dialog. An STA should be able to switch from the transmit mode to the receive mode within this specified period. Frames that use SIFS include acknowledgment (ACK) frames, CTS frames, data frame fragments of an excessively long MAC frame, and all AP probe response frames.

PCF interframe space (PIFS) is used only by STAs working in PCF mode. With this IFS, an STA can have the privilege to occupy a channel during the contention-free period (CFP). However, the PCF mechanism is beyond the scope of this book.

DCF interframe space (DIFS) is used by STAs working in DCF mode to transmit data frames and management frames. An STA, to transmit RTS frames or data frames, must continuously monitor an idle channel until the idle duration reaches the DIFS length. If the channel is busy, the transmission is delayed. After the channel is detected to be continuously idle for the DIFS period, an RB process needs to be started.

Arbitration interframe space (AIFS) is length-variable and introduced in the WLAN enhanced distributed channel access (EDCA) mechanism to preferentially transmit high-priority packets, unlike fixed-length DIFS that provides the same opportunities to all data packets. EDCA classifies data packets into four access categories (ACs) in descending order of priority. The AIFS also varies with the AC. AIFS[i] represents the time an STA needs to wait before starting a backoff process. A higher AC priority indicates a smaller AIFS[i], allowing an STA to start a backoff process earlier. This increases the probability of high-priority channels accessing the channel.

Based on their lengths, the abovementioned IFS types are prioritized as follows: AIFS[i] < DIFS < PIFS < SIFS.

2.2.6.3 RB

Multiple STAs waiting to transmit data on a busy channel may have identical NAVs. When the countdown reaches 0, the STAs may all consider the channel as idle and transmit data simultaneously. As wireless STAs cannot transmit data and listen to the channel state at the same time, the probability of such conflict is high. Therefore, to reduce the probability of conflicts, after STAs with data to be transmitted detect that a channel is busy, an RB mechanism needs to be started, allowing them to transmit data at different times. The implementation method is as follows: After detecting an idle channel, an STA continues channel listening for a period of time that is the IFS time and RB time combined (the RB time basically varies among STAs). Then, if the channel remains idle, the STA can occupy the channel. The STA with the shortest backoff time first transmits an RTS frame to occupy the channel. Then, the other STAs update their NAVs using the CS process, and they start another round of RB after the channel becomes idle again.

If an STA detects an idle channel for transmitting its first data frame, no RB is required.

The RB time, also referred to as a contention window, is an integer multiple of a timeslot. Its length is determined by the physical-layer technology. After an STA that wants to transmit data selects the RB time based on the backoff algorithm, a backoff timer is specified. In this time period, the STA senses the channel during every other timeslot, and the timer decreases by one timeslot when the channel is idle. The timer freezes when the channel is busy and resumes counting when the channel is idle again. Then, after the STA succeeds in channel contention and transmits data frames, a new round of backoff starts, as illustrated in Figure 2.34. This backoff algorithm is used to maintain fairness when STAs attempt to occupy channels. That is, an STA that fails to occupy a channel enters the next contention with a shorter backoff time, thereby avoiding constant channel contention failures.

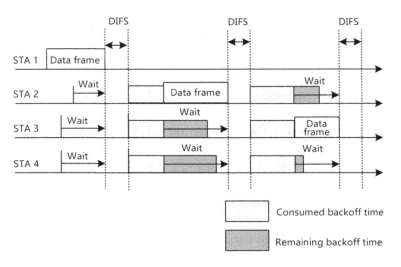

FIGURE 2.34   RB mechanism.

2.3 802.11 STANDARDS

This section describes the characteristics of standards in the 802.11 family and compares their differences.

2.3.1 Origin (802.11–1997)

In 1990, the Institute of Electrical and Electronics Engineers (IEEE) established the 802.11 working group to standardize WLAN, and following decades of development, 802.11 has evolved into an entire suite of standards.

The earliest developed version in the family, IEEE 802.11, also known as 802.11–1997, was completed in 1997, and it defines the MAC layer and the physical layer. It is preferred over complex technologies that use centralized access protocols (such as HiperLAN) and is used as the WLAN standard in the industry due to the following features:

- Ethernet-like simple access protocol

- Simple distributed access mechanism (CSMA/CA) to ensure communication quality

802.11–1997 defines one transmission mode (infrared spectrum) and two modulation modes (2.4 GHz FHSS and 2.4 GHz DSSS) at the physical layer.

2.3.2 Enhancement (802.11b and 802.11a)

The 802.11 working group formulated two revised versions of the standards in 1999:

- 802.11b uses DSSS to improve data transmission rates in the 2.4 GHz band (denoted as 802.11b–2.4 GHz) to up to 11 Mbps.

- 802.11a establishes a new physical layer in the 5 GHz frequency band (denoted as 802.11a–5 GHz). It delivers data transmission rates of up to 54 Mbps.

802.11b–2.4 GHz also uses complementary code keying (CCK) to enhance DSSS and deliver data transmission rates of up to 11 Mbps. Due to this advantage, 802.11b devices proved to be a huge market success when compared to IR and FHSS devices.

802.11a–5 GHz introduced OFDM technology into the 802.11 standards. However, despite the support for data transmission rates of up to 54 Mbps, 802.11a–5 GHz did not develop rapidly due to a limited number of channels available in the 5 GHz frequency band. In addition, as new devices need to be backward compatible with 802.11b devices, they also need to be configured with two wireless modules to run 802.11a–5 GHz and 802.11b–2.4 GHz separately.

2.3.3 Standard Extension and Compatibility (802.11g)

With permission from Federal Communications Commission (FCC) to use OFDM in the 2.4 GHz frequency band in 2001, the IEEE 802.11 working group formulated 802.11g in 2003 to apply OFDM in 802.11a to the 2.4 GHz frequency band.

802.11g supports backward compatibility and interoperability with 802.11b devices, allowing a new 802.11g STA to connect to an 802.11b AP, and an old 802.11b STA to connect to a new 802.11g AP, for example. 802.11g delivers data transmission rates of up to 54 Mbps. These advantages ensured that 802.11g products achieved great success in the market.

2.3.4 MIMO-OFDM-Based HT (802.11n)

In 2002, the IEEE 802.11 working group established a high throughput (HT) research group to develop a next-generation standard. Based on

MIMO-OFDM, it was officially released in 2009 as IEEE 802.11n with the following advantages:

- Supports a maximum of four spatial streams.

- Defines single-user beamforming to improve reception.

- Delivers a data transmission rate of up to 300 Mbps with the 20 MHz channel bandwidth or 600 Mbps with the 40 MHz channel bandwidth.

- Incorporates 802.11e to improve real-time service quality. More specifically, 802.11n devices must support 802.11e features.

2.3.5 VHT (802.11ac)

With the rapid growth of multimedia services and the subsequent desire for higher transmission rates, the IEEE 802.11 working group formally released the 802.11ac standard in 2014. This standard, also referred to as a very high throughput (VHT) standard, includes the following features:

- Increases the maximum number of spatial streams from four in 802.11n to eight.

- Increases the channel bandwidth from 40 to 160 MHz and delivers data transmission rates of up to 6933.3 Mbps.

- Defines downlink MU-MIMO to empower downlink multiuser parallel transmission.

2.3.6 HEW (802.11ax)

The next-generation standard 802.11ax, also referred to as high efficiency WLAN (HEW), includes the following features:

- Uses OFDMA to allow for narrower subcarrier spacing and improved data transmission robustness and throughput.

- Introduces uplink MU-MIMO to further improve the throughput and service quality in high-density scenarios.

2.3.7 Comparisons between 802.11 Standards

Table 2.3 lists the differences between 802.11 standards.

TABLE 2.3 Comparisons between 802.11 Standards

Standard Version	Year Issued	Frequency (GHz)	Physical-Layer Technology	Modulation Mode	Number of Spatial Streams	Channel Bandwidth (MHz)	Data Rate (Mbps)
802.11	1997	2.4	IR, FHSS, DSSS	-	-	20	1 and 2
802.11b	1999	2.4	DSSS, CCK	-	-	20	5.5 and 11
802.11a	1999	5	OFDM	64QAM	-	20	6–54
802.11g	2003	2.4	OFDM, DSSS, CCK	64QAM	-	20	1–54
802.11n (HT)	2009	2.4, 5	OFDM, SU-MIMO	64QAM	4	20, 40	6–600
802.11ac (VHT)	2014	5	OFDM, downlink MU-MIMO	256QAM	8	20, 40, 80, 160, 80+80	6–6933.3
802.11ax (HEW)	2020 (estimated)	2.4, 5, 6	OFDMA, downlink MU-MIMO, uplink MU-MIMO	1024QAM	8	20, 40, 80, 160, 80+80	6–9607.8

2.4 802.11AX STANDARDS

Each generation, from 802.11 to 802.11ac, features an improved transmission rate at the physical layer. However, not every generation has significantly improved spectral efficiency. As a result, the average user rate cannot significantly increase in high-density WLAN deployment scenarios. Due to the scarcity of spectrum resources and complexity of devices, WLANs have experienced limited performance improvement by increasing the number of bonded channels and the number of MIMO spatial streams. In addition, the contention-based random access mechanism, by wasting large amounts of channel resources, generates low spectrum utilization. Therefore, to improve WLAN performance, the most effective measure is to improve spectral efficiency and change the access mechanism.

As such, 802.11ax, also referred to as HEW, introduces key technologies such as OFDMA and uplink MU-MIMO to improve spectrum utilization. This standard also optimizes the throughput, anti-interference capability, and energy efficiency. The Wi-Fi Alliance named this standard Wi-Fi 6 (i.e., the sixth generation of WLAN).

2.4.1 802.11ax Overview

A member of the 802.11 family, 802.11ax features multiple improvements on the physical layer and MAC layer:

- Excellent compatibility with 802.11a, 802.11b, 802.11g, 802.11n, and 802.11ac

- Improved spectrum utilization, a fourfold increase in actual user throughput in high-density WLAN deployment scenarios

- High throughput of up to 9607.8 Mbps

- Improved anti-interference capabilities, especially in outdoor transmission scenarios

- Energy efficiency, increasing the battery life of terminals

2.4.1.1 Physical-Layer Technology of 802.11ax

The development of physical-layer technologies is essential to the evolution of wireless networks. 802.11ax introduces multiple technologies to achieve physical-layer enhancement, including coding, modulation, multiuser, channel bonding, and a new PPDU format to improve WLAN throughput, anti-interference capability, and spectrum utilization.

1. Coding

   Coding is used to reduce the bit error rate. Instead of introducing new coding schemes, 802.11ax uses BCC and LDPC, but it clearly specifies their application scenarios to ensure that the optimal coding scheme is used.

2. Modulation scheme

   The selected modulation scheme affects the WLAN throughput and anti-interference capability. 802.11ax introduces high-order modulation 1024QAM to carry more information throughout the same time period. In addition, 802.11ax divides a channel into more subcarriers to carry more information over the same channel bandwidth and time period and improve its anti-interference capability during long-distance transmission. As using higher-order modulation in an environment with strong interference may not deliver desired performance, 802.11ax also introduces dual-carrier modulation (DCM) to modulate the same information on a pair of subcarriers, improving anti-interference capability.

3. Multiuser technology

   The multiuser technology can improve spectrum utilization. 802.11ax introduces OFDMA to divide a channel into RUs that can be allocated to different users and to flexibly adjust multiuser concurrent transmission. OFDMA can also be used with MU-MIMO to further improve spectrum utilization.

4. Channel bonding

   The channel bandwidth affects transmission throughput of terminals. 802.11ax supports bonding of noncontiguous channels rather than larger-bandwidth channels. When a secondary channel experiences interference, other channels are still bonded to maintain high-speed transmission, improving the anti-interference capability of channel bonding.

5. PPDU

   PPDU is a data packet format at the physical layer. 802.11ax defines four efficient data packet formats and provides compatibility with earlier 802.11 standards to achieve efficient transmission.

2.4.1.2 MAC-Layer Technology of 802.11ax

The MAC layer specifies the transmission mode of data frames on a channel. 802.11ax introduces multiple new MAC-layer technologies to improve WLAN transmission efficiency and battery life, including multiuser transmission, energy efficiency, and spatial multiplexing.

1. Multiuser transmission

   802.11ax implements multiuser transmission in OFDMA and MU-MIMO to improve the uplink and downlink transmission mechanism, allowing APs to flexibly and efficiently schedule multiple terminals to perform concurrent uplink transmission. In doing this, the probability of a collision occurring when terminals contend for a channel in high-density scenarios is significantly reduced, while data transmission efficiency is also improved.

2. Energy efficiency

   802.11ax defines a target wake-up time (TWT) in situations where multiple users concurrently perform transmission.

   A TWT mechanism is used to improve the transmission efficiency of an STA that periodically transmits and receives small amounts of data, while also reducing the power consumption of the STA.

3. Spatial multiplexing

   802.11ax uses spatial multiplexing to address the interference caused by channel overlapping in dense AP deployment scenarios. When specific conditions are satisfied, spatial multiplexing allows data to be transmitted concurrently, thereby improving transmission efficiency.

2.4.2 High Efficiency PPDU

High efficiency PPDUs are used to enhance the 802.11ax physical layer. This section starts by describing four new PPDU formats. Figure 2.35 illustrates the four PPDU formats.

- High efficiency single-user PPDU (HE SU PPDU) format: This defines the packet format for communications between an AP and an STA and communications between two STAs.

- High efficiency multiuser PPDU (HE MU PPDU) format: This defines concurrent transmission among multiple users.

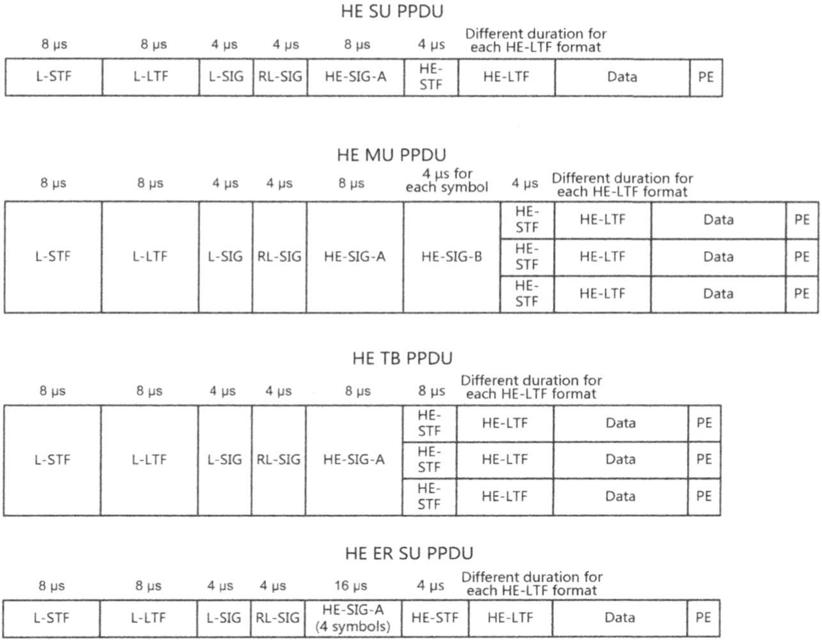

HE SU PPDU

8 µs	8 µs	4 µs	4 µs	8 µs	4 µs	Different duration for each HE-LTF format		
L-STF	L-LTF	L-SIG	RL-SIG	HE-SIG-A	HE-STF	HE-LTF	Data	PE

HE MU PPDU

8 µs	8 µs	4 µs	4 µs	8 µs	4 µs for each symbol	4 µs	Different duration for each HE-LTF format		
						HE-STF	HE-LTF	Data	PE
L-STF	L-LTF	L-SIG	RL-SIG	HE-SIG-A	HE-SIG-B	HE-STF	HE-LTF	Data	PE
						HE-STF	HE-LTF	Data	PE

HE TB PPDU

8 µs	8 µs	4 µs	4 µs	8 µs	8 µs	Different duration for each HE-LTF format		
					HE-STF	HE-LTF	Data	PE
L-STF	L-LTF	L-SIG	RL-SIG	HE-SIG-A	HE-STF	HE-LTF	Data	PE
					HE-STF	HE-LTF	Data	PE

HE ER SU PPDU

8 µs	8 µs	4 µs	4 µs	16 µs	4 µs	Different duration for each HE-LTF format		
L-STF	L-LTF	L-SIG	RL-SIG	HE-SIG-A (4 symbols)	HE-STF	HE-LTF	Data	PE

FIGURE 2.35   Four PPDU formats.

- High efficiency trigger-based PPDU (HE TB PPDU) format: This format defines OFDMA or MU-MIMO uplink data transmission. The trigger frame contains resource allocation information used for concurrent uplink data transmissions upon receiving of the trigger frame. The uplink multiuser transmission format is also known as the trigger-based format.

- High efficiency extended range single-user PPDU (HE ER SU PPDU) format: This is an extension of the HE SU PPDU, introduced in 802.11ax to improve outdoor transmission performance. For a preamble, high efficiency signal field A (HE-SIG-A) is repeated (occupying an increased number of symbols, specifically, from two in other formats to four) to improve transmission reliability. Moreover, power boosting (PB) is applied to legacy preambles, high efficiency short training fields (HE-STFs), and high efficiency long training fields (HE-LTFs). This is done to extend transmission ranges. For the data field, its transmission range is extended using DCM and 106-tone RU narrowband transmission.

An 802.11ax-compliant PPDU packet (or HE PPDU) consists of the preamble, data field, and packet extension (PE).

- A preamble is a string of repeated 0101 for information synchronization such as clock information between a transmitter and a receiver, for the transmission of data fields. The preamble of an HE PPDU includes pre-HE modulated fields and HE modulated fields. These fields will be described later in this section.

- A data field carries payload.

- A PE spares a receiver more processing time in receiving an HE PPDU. To balance PPDU processing time and overheads for STAs, 802.11ax defines four PE durations: 0, 4, 8, and 16 μs.

2.4.2.1 Pre-HE Modulated Fields

A pre-HE modulated field consists of the legacy preamble field, repeated legacy signal field (RL-SIG), HE-SIG-A, and high efficiency signal field B (HE-SIG-B). The fields differ in different types of PPDUs. For example, the HE-SIG-B field is exclusive for HE MU PPDUs, and the HE-SIG-A field in an HE ER SU PPDU is extended to four symbols.

The non-HE portion consists of a legacy preamble in 802.11ax. The remaining portion of a PPDU apart from the non-HE portion is referred to as the HE portion. Figure 2.36 illustrates the HE MU PPDU format.

1. Non-HE portion (legacy preamble)

    The non-HE portion includes a legacy short training field (L-STF), a legacy long training field (L-LTF), and a legacy signal field (L-SIG). Table 2.4 describes the functions of these fields.

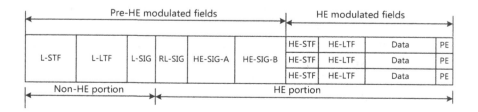

FIGURE 2.36   HE MU PPDU format.

TABLE 2.4   Functions of the Non-HE Fields

Field	Description	Transmission Duration (μs)	New Features
L-STF	The L-STF field contains ten repetitions of a portion for receivers to identify PPDUs and obtain information about (and compensate for) frequency and time	8	Beam change, PB
L-LTF	The L-LTF field contains two repetitions of a 3.2 μs portion and one 1.6 μs GI. According to the L-LTF field, receivers further obtain information about (and compensate for) frequency and time, and perform channel estimation (including SNR estimation)	8	Beam change, PB
L-SIG	The L-SIG field carries rate and length information, which indicates the duration of the PPDU. Transmission needs to be delayed by at least the duration of the PPDU for a conventional STA, to avoid interference. Both HE STAs and conventional STAs can decode the L-SIG field in the non-HE portion for information. This further ensures backward compatibility	4	Beam change, PB, length indivisible by 3, four extra subcarriers

The non-HE portion distinguishes itself from the ones of previous standards in the following features:

- Beam change: 802.11ax introduces a beam change mechanism. The L-STF field is transmitted using a mechanism similar to the L-STF transmission mechanism in 802.11n and 802.11ac when BEAM_CHANGE for the transmit vector is 1 or is not present. If multiple antennas are used in a transmission system, cyclic shift is performed. A similar beam change mechanism applies to the L-LTF and L-SIG fields. The same spatial mapping as that of HE modulated fields shall be used for L-STF, L-LTF, and L-SIG when BEAM_CHANGE for the transmit vector is 0. Particularly, with L-LTF, receivers can enhance channel estimation provided by the HE-LTF field based on channel estimation provided by the L-LTF field. This thereby enhances the performance of demodulating the data portion.

- PB: In an HE ER SU PPDU, the power of L-STF, L-LTF, and L-SIG is boosted by 3 dB to further extend the transmission range of the HE ER SU PPDU.

- Length indivisible by 3: The length of the L-SIG field is indivisible by 3, which identifies HE PPDUs. The length of the L-SIG field is always a multiple of 3 in both an HT PPDU and a VHT PPDU in earlier standards, whereas this multiple of 3 is further reduced by 1 or 2 in an HE PPDU. An HE MU PPDU or an HE ER SU PPDU has 1 subtracted, and the rest of the formats of HE PPDU have 2 subtracted. For more information on the identification process, see the descriptions of autodetection in this section.

- Four extra subcarriers: In standards earlier than 802.11ax, the L-SIG field is spread over 48 data subcarriers and four pilot subcarriers as defined, meaning this field is carried on 52 subcarriers among a total of 64 subcarriers. In 802.11ax, two subcarriers are added at each edge of the previous subcarriers for channel estimation. This means another four subcarriers are added. This further improves spectrum utilization of the signal field, together with the four subcarriers and the 52 subcarriers for the L-LTF field, which are utilized for channel estimation. In this case, channel estimation is performed on 56 subcarriers in total. With channel estimation over 56 subcarriers, the HE-SIG-A and HE-SIG-B fields following the L-SIG field have four more subcarriers for carrying information in each symbol than those in earlier standards. As a result, this improves spectrum utilization.

2. RL-SIG

   The RL-SIG field is a repetition of the L-SIG field and shares all features of the L-SIG field. The RL-SIG field is exclusive for HE PPDUs. The L-SIG repetition brings the following advantages:

   - Reliability is enhanced. After symbols are repeated, maximum ratio combining (MRC) is performed at receivers. In this manner, the equivalent SNR is boosted by 3 dB, enhancing HE PPDU reliability, especially for outdoor transmissions. MRC is a technology that improves receive signal quality through algorithms.

   - The HE PPDU format can be differentiated from other PPDU formats through autodetection. Receivers detect whether the L-SIG field is similar to the following symbol, and the result can be used to identify an HE PPDU in the received PPDUs.

3. HE-SIG-A

The HE-SIG-A field carries information for helping parse HE PPDUs.

HE PPDUs mentioned above can be displayed in four different formats, and carried information and the length of the HE-SIG-A field vary with HE PPDU formats.

• In the HE SU PPDU, HE MU PPDU, and HE TB PPDU formats, the HE-SIG-A field spreads over two symbols in total, namely, HE-SIG-A1 and HE-SIG-A2.

• In the HE ER SU PPDU format, the HE-SIG-A field spreads over four symbols in total, namely HE-SIG-A1, HE-SIG-A2, HE-SIG-A3, and HE-SIG-A4, with the former two symbols carrying the same content and the latter two symbols carrying the same content.

The following describes the content of HE-SIG-A in different formats in detail.

Figure 2.37 illustrates the content of HE-SIG-A in the HE SU PPDU and HE ER SU PPDU formats.

Table 2.5 describes the common functions of fields in HE-SIG-A.

Figure 2.38 illustrates the content of HE-SIG-A in the HE MU PPDU format.

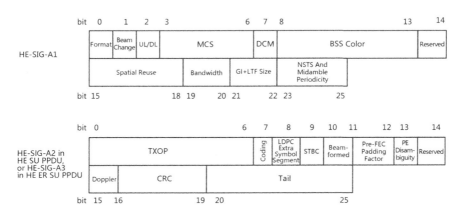

FIGURE 2.37   Content of HE-SIG-A in the HE SU PPDU and HE ER SU PPDU formats.

TABLE 2.5   Functions of Common Fields in HE-SIG-A

Field	Number of Bits	Description
Format	1	Used to differentiate HE SU PPDU or HE ER SU PPDU from HE TB PPDU If it is set to 1, the PPDU format is HE SU PPDU or HE ER SU PPDU
Beam change	1	Indicates whether pre-HE modulated fields and HE modulated fields use the same spatial mapping
DCM	1	Indicates whether DCM is performed
BSS color	6	Used to identify a basic service set (BSS)
Spatial reuse	4	Indicates the parameters required for spatial multiplexing
GI+LTF size	2	Indicates the combination of the GI and HE-LTF length
NSTS and midamble periodicity	3	Indicates the numbers of space-time streams in non-Doppler mode, or both the number of space-time streams and midamble period in Doppler mode
TXOP	7	Indicates the transmission opportunity duration based on which the time of using the air interface is reserved
Doppler	1	Indicates whether the Doppler mode is used

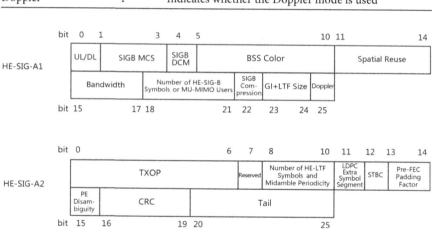

FIGURE 2.38   Content of the HE-SIG-A field in an HE MU PPDU.

The HE MU PPDU format provides HE-SIG-B indications in HE-SIG-A, which the HE SU PPDU and HE ER SU PPDU formats do not provide. These indications include the number of HE-SIG-B symbols or the number of MU-MIMO users, HE-SIG-B MCS, and the HE-SIG-B compression mode (which indicates whether to perform compression).

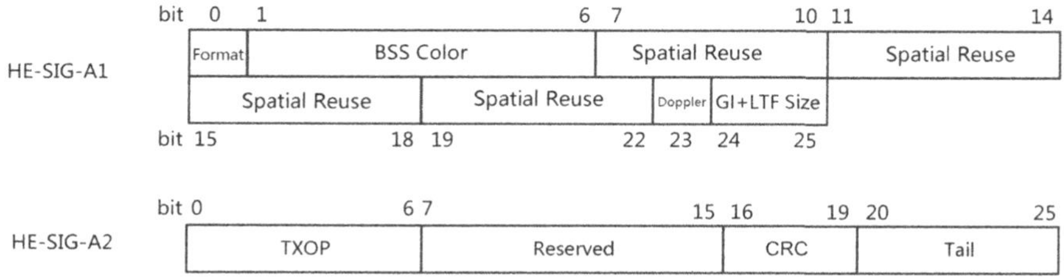

bit 0   1   6 7   10 11   14

| Format | BSS Color | Spatial Reuse | Spatial Reuse |

HE-SIG-A1

| Spatial Reuse | Spatial Reuse | Doppler | GI+LTF Size |

bit 15   18 19   22 23 24   25

bit 0   6 7   15 16   19 20   25

HE-SIG-A2

| TXOP | Reserved | CRC | Tail |

FIGURE 2.39   Content of HE-SIG-A in an HE TB PPDU.

Modulation and coding parameters for the data portion are found in the HE-SIG-B field and are specific for different users.

Figure 2.39 illustrates the content of HE-SIG-A in an HE TB PPDU. The HE-SIG-A field in an HE TB PPDU carries the content as indicated in the trigger information sent from an AP. Therefore, the AP does not need to read the content of the HE-SIG-A field, but some content in this field can be read by other APs and STAs. The number of fields using spatial multiplexing is increased to enhance spatial multiplexing, because the majority of the information in an HE TB PPDU does not need to be specified. For details related to spatial multiplexing, see Section 2.4.5. A beam change mechanism is also applied on the HE-SIG-A field in an HE PPDU, similar to that for the L-STF, L-LTF, L-SIG, and RL-SIG fields. Details are not described herein.

The following describes the modulation and coding of HE-SIG-A. Two MCSs are available for HE-SIG-A:

- BCC BPSK modulation is used in an HE SU PPDU, HE MU PPDU, or HE TB PPDU.

- HE-SIG-A occupies four symbols and uses BCC modulation in an HE ER SU PPDU. HE-SIG-A1 and HE-SIG-A3 use BPSK modulation. Unlike HE-SIG-A1, HE-SIG-A2 uses quadrature binary phase shift keying (QBPSK) modulation without interleaving. HE-SIG-A4 also uses BPSK modulation. However, unlike HE-SIG-A3, HE-SIG-A4 does not use interleaving.

4. HE-SIG-B

HE-SIG-B provides resource allocation information for OFDMA and MU-MIMO. Figure 2.40 illustrates the structure of the HE-SIG-B on each 20 MHz channel.

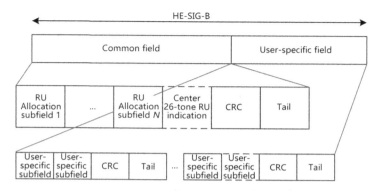

FIGURE 2.40   Structure of the HE-SIG-B field on each 20 MHz channel.

HE-SIG-B has two portions.

- Common field: includes 1 to *N* RU allocation subfields and a sub-field for center 26-tone RU indication if the bandwidth is greater than or equal to 80 MHz. It also includes a cyclic redundancy check (CRC) subfield for checks, and a tail subfield for cyclic decoding. *N* depends on the bandwidth. Specifically, if the band-width is 20 or 40 MHz, *N* is 1. If the bandwidth is 80 MHz, *N* is 2 and if the bandwidth is 160 MHz, *N* is 4.

- User-specific field: consists of 1 to *M* user-specific subfields ordered in the RU allocation sequence. Two user-specific sub-fields are grouped together in most cases (except for the last group), and every two user-specific subfields are followed by a CRC subfield and a tail subfield. The last group of user-specific subfields may include one or two user-specific subfields, a CRC subfield and a tail subfield.

The common field provides RU allocation information. The follow-ing describes an RU allocation indication.

802.11ax introduces the concept of content channel (CC). If 20 MHz bandwidth is available for data packet transmission, the HE-SIG-B field includes only one CC. If 40 MHz bandwidth is avail-able for data packet transmission, the HE-SIG-B field includes CC 1 and CC 2. If 80 MHz bandwidth is available for data packet trans-mission, the HE-SIG-B field still includes CC 1 and CC 2. In this case, there are a total of four 20 MHz channels, and RU allocation

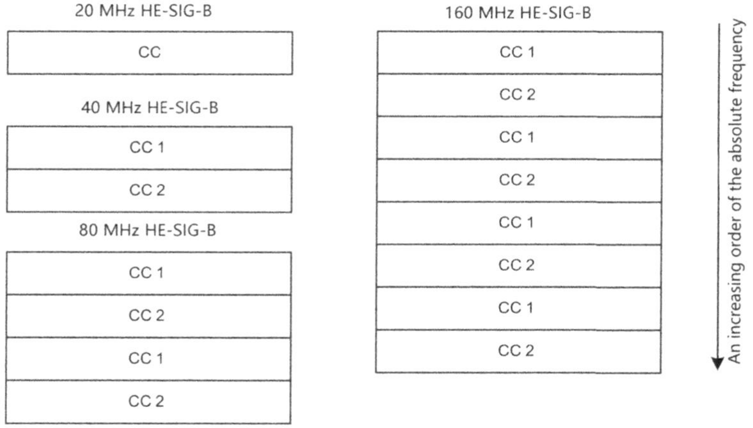

20 MHz HE-SIG-B

CC

40 MHz HE-SIG-B

CC 1
CC 2

80 MHz HE-SIG-B

CC 1
CC 2
CC 1
CC 2

160 MHz HE-SIG-B

CC 1
CC 2
CC 1
CC 2
CC 1
CC 2
CC 1
CC 2

An increasing order of the absolute frequency

FIGURE 2.41   CCs in different bandwidths.

information is available on the four channels in the sequence CC 1, CC 2, CC 1, and CC 2 (in an increasing order of the absolute frequency). If 160 MHz bandwidth is available for data packet transmission, expansion is performed in the same manner as in 80 MHz bandwidth, as illustrated in Figure 2.41.

If 20 MHz bandwidth is available for data packet transmission, the CC includes one RU allocation subfield, which indicates how the 242-tone RU is allocated. For more information on RU allocation principles, see Section 2.4.4. Correspondingly, the user-specific field contains information for all STAs designated by the 242-tone RU allocation subfield in the allocation sequence. The 8 bit RU allocation subfield indicates all possible RU combinations within the 242-tone RU by respective indexes. In addition, for an RU greater than or equal to 106-tone, an index serves as the number of users using SU-MIMO or MU-MIMO transmission with the RU. As illustrated in Figure 2.42, the RU allocation subfield is 01000010 and the index table clearly shows that a 242-tone RU is allocated into 106-, 26-, 26-, 26-, 26-, and 26-tone, and 26-tone RUs with the 106-tone RU serving three users.

The transmission is performed on CC 1 and CC 2 if 40 MHz bandwidth is available. CC 1 contains RU allocation subfields and corresponding user-specific fields in the first 242-tone RU and CC 2 contains those in the second 242-tone RU.

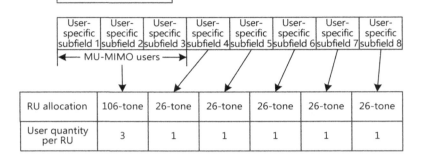

FIGURE 2.42 Mapping between RU allocation subfields and user-specific fields.

If 80 MHz bandwidth is available for data packet transmission, CCs carry content as illustrated in Figure 2.43. Both CCs carry center 26-tone RU indication in 80 MHz to indicate whether the corresponding RU is used to transmit data.

If 160 MHz bandwidth is available for data packet transmission, CC 1 contains RU allocation subfields within the first, third, fifth, and seventh 242-tone RUs and corresponding user-specific fields. CC 2 contains RU allocation subfields within the second, fourth, sixth,

CC	RU allocation subfield within the first 242-tone RU	RU allocation subfield within the third 242-tone RU	Center 26-tone RU indication	CRC+tail	Corresponding user-specific fields in the first and third 242-tone RU
CC	RU allocation subfield within the second 242-tone RU	RU allocation subfield within the fourth 242-tone RU	Center 26-tone RU indication	CRC+tail	Corresponding user-specific fields in the second and fourth 242-tone RU
CC	RU allocation subfield within the first 242-tone RU	RU allocation subfield within the third 242-tone RU	Center 26-tone RU indication	CRC+tail	Corresponding user-specific fields in the first and third 242-tone RU
CC	RU allocation subfield within the second 242-tone RU	RU allocation subfield within the fourth 242-tone RU	Center 26-tone RU indication	CRC+tail	Corresponding user-specific fields in the second and fourth 242-tone RU

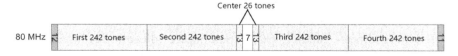

FIGURE 2.43 HE-SIG-B indication in 80 MHz.

and eighth 242-tone RUs and corresponding user-specific fields. In addition, the two CCs respectively carry center 26-tone RU indication of the first 80 MHz and the second 80 MHz, to indicate whether the corresponding RU is used to transmit data.

RU allocation subfields can also indicate an RU larger than 242-tone is used, if more than 20 MHz bandwidth is available for data packet transmission. For example, a 484-tone RU or 996-tone RU indicates that a larger RU including the 242-tone RU is allocated to an STA.

For user-specific fields, non-MU-MIMO RUs and MU-MIMO RUs contain different information. For non-MU-MIMO RUs, user-specific fields signal identifiers, numbers of space-time streams, beamforming completion indication, MCS, and DCM and coding indications for intended STAs. For MU-MIMO RUs, user-specific fields signal identifiers, spatial configurations, MCS, and reservation and coding indications for intended STAs.

The following special mechanisms are further introduced for the HE-SIG-B:

- In the case of full-bandwidth allocation for MU-MIMO, meaning only one RU on the entire bandwidth, HE-SIG-A indicates whether HE-SIG-B is compressed. HE-SIG-B includes only user-specific fields, but no common fields. The number of STAs participating in MU-MIMO can also be found in HE-SIG-A.

- Some special fields may be used to indicate that an RU does not serve users, when some RUs are not allocated to STAs, or a preamble is punctured. This leaves no need for such indications in a user-specific field.

- User-specific content fields are flexibly ordered across CC 1 and CC 2 by using some special indications, for MU-MIMO allocations of RU sizes larger than 242 tones. This thereby strikes a balance between lengths of CC 1 and CC 2 and reduces overheads.

5. Autodetection

WLAN devices are backward compatible. For example, 802.11ax-compliant devices can send not only HE PPDUs but also VHT PPDUs, HT PPDUs, and legacy PPDUs. Receivers can correspondingly identify and receive PPDUs of different formats.

Autodetection is a method for receivers to identify different types of PPDUs. 802.11ax-compliant HE PPDUs provide the following features to help receivers perform autodetection:

- Repetition of L-SIG and RL-SIG

- L-SIG length indivisible by 3

- L-SIG rate of 6 Mbps

As mentioned above, receivers need to further identify the four HE PPDU formats. As illustrated in Figure 2.44, the remainder (1 or 2) after the division of the L-SIG length by 3 differentiates HE SU PPDU/HE TB PPDU from HE MU PPDU/HE ER SU PPDU. A format identifier (bit 0) in the HE-SIG-A is used to further differentiate between an HE SU PPDU and an HE TB PPDU. The phase (BSPK/QBPSK) of a second symbol of HE-SIG-A is used to further differentiate between an HE MU PPDU and an HE ER SU PPDU.

2.4.2.2 HE Modulated Fields
HE modulated fields include HE-STF and HE-LTF.

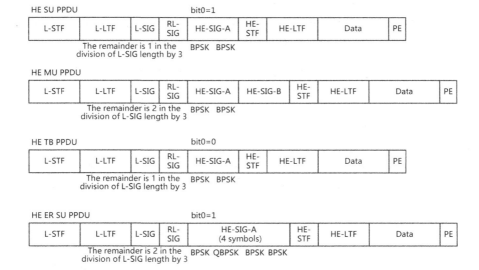

FIGURE 2.44   Differences in four PPDU formats.

1. HE-STF

   HE-STF aims to improve automatic gain control (AGC) esti-mation in MIMO operation. HE-STF is different from the high throughput short training field (HT-STF) and very high throughput short training field (VHT-STF) in the following aspects:

   - HE-STF is transmitted in units of RUs due to OFDMA operation.

   - There exist two durations. HE-STF has the same duration as HT-STF and VHT-STF in the HE SU PPDU, HE ER SU PPDU, and HE MU PPDU formats (4 μs, with 5 periods of 0.8 μs). The duration of HE-STF is 8 μs (with 5 periods of 1.6 μs) in the HE TB PPDU format. If the HE TB PPDU format is used, HE-STF may be transmitted only on one narrowband 26-tone RU. If the 4-μs HE-STF is used, the 26-tone RU may contain no HE-STF subcar-riers with energy. However, an 8-μs HE-STF with 5 periods of 1.6 μs ensures that each RU contains HE-STF subcarriers with energy, creating denser HE-STF subcarriers with energy. In addi-tion, a longer HE-STF enhances AGC performance.

2. HE-LTF

   HE-LTF aims to help receivers with channel estimation in MIMO operation. HE-LTF is different from high throughput long training field (HT-LTF) and very high throughput long training field (VHT-LTF) in the following aspects:

   - HE-LTF is transmitted in units of RUs due to OFDMA opera-tion. The PAPR of HE-LTF transmissions over narrowband RUs is considered into the HE-LTF sequence design.

   - HE-LTF transmissions are aligned in OFDMA operation. Different RUs may have different numbers of spatial streams in OFDMA operation. If this is the case, the required HE-LTF sym-bol quantities are also different. For this reason, some RUs may transmit additional HE-LTF symbols to align HE-LTF transmis-sions among RUs.

   - Various HE-LTF types are introduced in 802.11ax to balance per-formance and overheads for channel estimation. The types include 1x HE-LTF, 2x HE-LTF, and 4x HE-LTF, which have a respective duration of 3.2, 6.4, 12.8 µs, without GI. GI can be 0.8, 1.6, or 3.2 µs. For more information, refer to the 802.11ax standard.

- Various HE-LTF modes are introduced in 802.11ax, different from VHT-LTF that always uses the single stream pilot mode as defined in 802.11ac.

  - Single stream pilot HE-LTF mode: Single stream pilot in LTF is predominantly applied to HE SU, HE UL/DL OFDMA, DL MU-MIMO, and UL MU-MIMO not using 1x HE-LTF. Single stream pilot is a pilot mode in which different spatial streams use the same pilot sequence.

  - HE masked HE-LTF sequence mode: In this HE-LTF mode, HE-LTF sequences are masked, meaning an HE-LTF sequence in a symbol is multiplied by an orthogonal sequence. This mode is predominantly applied to UL MU-MIMO not using 1x HE-LTF. This mode enables a receiver (AP) to estimate frequency offset and compensate for frequency offset errors when receiving HE-LTFs during a UL MU-MIMO process.

  - 1x HE-LTF mode for non-OFDMA UL MU-MIMO: This mode uses neither single stream pilots nor mask sequences, which is different from the preceding two modes. This is because 1x HE-LTF has a relatively short duration and performance is slightly impacted even if frequency offset is not estimated and frequency offset errors are not compensated for.

  - HE-LTF mode in the HE TB null data packet (NDP) feedback PPDU format: This mode is different from the preceding three modes.

2.4.3 Throughput Improvement

802.11ax enhances modulation schemes and subcarrier division for OFDM operation. It increases the number of bits per symbol, the percentage of data subcarriers, and the percentage of time spent transmitting data. Specifically, the physical rate increases by 38.5% and throughput increases from 6933.3 to 9607.8 Mbps compared with those in 802.11ac.

Channel bonding is improved via preamble puncturing in 802.11ax. Other channels remain bonded so that high throughput is retained when bonded secondary channels experience interference.

2.4.3.1 High-Order Modulation Scheme

802.11ax defines 1024QAM to accommodate improvements to device performance and demodulation algorithms. This provides 10 bits in each modulated symbol and increases the peak data rate by 25% compared with 256QAM defined in 802.11ac.

2.4.3.2 OFDM Subcarrier Division

In OFDM operation in 802.11ac, a 20 MHz channel is divided into 64 subcarriers, 56 of which are working subcarriers (known as pilot subcarriers and data subcarriers), accounting for 87.5%. 802.11ax introduces a finer granularity on subcarrier division. Specifically, each 20 MHz channel consists of 256 subcarriers. As illustrated in Figure 2.45, there are 234 data subcarriers, accounting for 91.4% of the total. In addition, thanks to the channel bonding technology, for example, on an 80 MHz channel, the peak data rate in 802.11ax increases by 4.7% compared with that in 802.11ac.

An approximate four-fold increase in the subcarrier quantity means a four-fold reduction in subcarrier spacing, that is, the frequency bandwidth of each subcarrier is reduced to 78.125 kHz. Given unchanged overall power, the duration for transmitting one symbol per subcarrier, known as symbol length, also increases four-fold. This in turn requires longer inter-symbol GIs to reduce the probability of inter-symbol interference. 802.11ac introduces a standard symbol length of 3.2 µs and a shortest GI of 0.4 µs. The percentage of time used for data transmission accounts for 88.9%. In contrast, 802.11ax introduces a standard symbol length of 12.8 µs and a shortest GI of 0.8 µs. The percentage of time used for data transmission tops at 94%. The peak data rate in 802.11ax increases by 5.88% compared with that in 802.11ac.

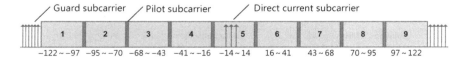

FIGURE 2.45   Subcarrier division in 802.11ax.

2.4.3.3 Preamble Puncturing

The channel bandwidth is 20 MHz in 802.11a and 802.11g. Channel bonding is introduced in 802.11n to allow 20 and 40 MHz channel bandwidths, and then extends further with 802.11ac to allow 20 MHz, 40 MHz, 80 MHz, 160 MHz, and 80 MHz+80 MHz channel bandwidths. Channel bonding increases the maximum channel bandwidth supported in each generation of Wi-Fi standards and boosts system throughputs accordingly.

However, contiguous channel bonding (CCB) also has drawbacks. For example, when a narrowband secondary channel among 802.11ac bonded channels is busy, the transmitter cannot use a secondary channel with a larger bandwidth. As illustrated in Figure 2.46, on the 802.11ac 80 MHz channel, if a secondary 20 MHz channel is busy, even if the secondary 40 MHz channel is idle, the transmitter can use only the primary 20 MHz channel. This wastes air interface resources of the secondary 40 MHz channel. In high-density scenarios, some channels are busy, which further increases the probability of channel fragmentation.

To address this issue, 802.11ax retains legacy 802.11ac standards (like CCB) and introduces preamble puncturing to further improve spectrum utilization. Preamble puncturing during initial discussion is referred to as noncontiguous channel bonding (NCB).

Preamble puncturing patterns are introduced for HE MU PPDU in 802.11ax. The HE-SIG-A field in the preamble of an HE MU PPDU includes a 3-bit bandwidth field which can be set to eight different values (**0–7**). There are four basic non-preamble puncturing patterns where the bandwidth field can be set to the following:

0 for 20 MHz; **1** for 40 MHz; **2** for non-preamble punctured 80 MHz; **3** for non-preamble punctured 160 MHz and 80 MHz+80 MHz.

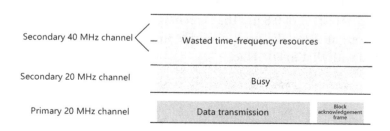

FIGURE 2.46 Possible channel fragmentation caused by channel bonding in 802.11ac.

Data packet bandwidth modes corresponding to the four patterns supported by VHT PPDU in 802.11ac are the same as those supported by HE SU PPDU and HE TB PPDU in 802.11ax. The following describes four preamble puncturing patterns specific to HE MU PPDU where the bandwidth field can be set to:

4 for preamble punctured 80 MHz. In this case, only the preamble of the secondary 20 MHz channel is punctured, as illustrated in Figure 2.47. Under this condition, even if the secondary 20 MHz channel is busy, APs may still send data by using spectrum resources of the primary 20 MHz channel and the secondary 40 MHz channel. Nevertheless, in non-preamble puncturing, only the primary 20 MHz channel can be availed. Therefore, preamble puncturing achieves three times spectrum utilization than non-preamble puncturing.

5 for preamble punctured 80 MHz. In this case, only the preamble of one of the two 20 MHz channels in the secondary 40 MHz channel is punctured, as illustrated in Figure 2.48.

As illustrated in the figure above, two different puncturing patterns are present under this condition. However, the receiver does not bother to distinguish between the two patterns because the corresponding resource allocation information is contained in signal fields on the primary and secondary 20 MHz channels. The receiver may further know which RUs are allocated by the APs through the RU allocation subfield of the HE-SIG-B.

6 for preamble punctured 160 MHz or 80 MHz+80 MHz. In this case, only the preamble of the secondary 20 MHz channel is punctured in the primary 80 MHz channel. Additionally, only the puncturing of the primary 80 MHz channel is specified. As for the secondary 80 MHz channel, any channel can be either punctured or not punctured, as illustrated in Figure 2.49.

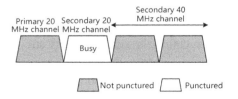

FIGURE 2.47 Preamble puncturing pattern with the bandwidth field set to 4.

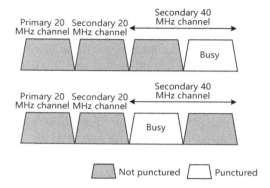

FIGURE 2.48   Preamble puncturing pattern with the bandwidth field set to 5.

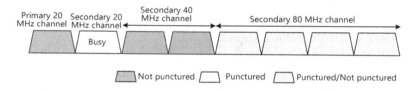

FIGURE 2.49   Preamble puncturing pattern with the bandwidth field set to 6.

7 for preamble punctured 160 MHz or 80 MHz+80 MHz. In this case, the primary 40 MHz channel is not punctured in the primary 80 MHz channel. As illustrated in Figure 2.50, three bandwidth combinations are available for the secondary 40 MHz channel in the primary 80 MHz. For the secondary 80 MHz channel, any channel can be either punctured or not punctured.

Similar to non-preamble puncturing, preamble puncturing for HE MU PPDUs in 802.11ax specifies that a single RU is still allocated to each STA, thereby greatly reducing the complexity of function implementation at the receiver.

Preamble puncturing significantly improves spectrum utilization and is most suited to radar channels and channels subject to narrowband interference.

In addition to the HE MU PPDU, preamble puncturing is also designed specifically for NDPs which are used for channel measurement in 802.11ax so that APs can learn states of non-contiguous channels.

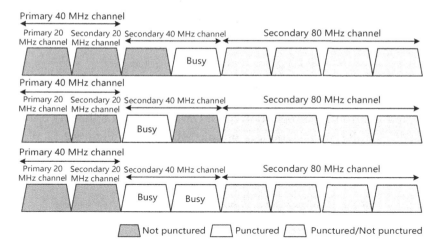

FIGURE 2.50   Preamble puncturing pattern with the bandwidth field set to 7.

2.4.4 Novel Multiuser Features

2.4.4.1 Introduction to Multiuser Features

In legacy Wi-Fi systems, including those in compliance with IEEE 802.11a, 802.11n, and 802.11ac, only point-to-point data transmissions are supported in the uplink. 802.11ax introduces OFDMA, which is a new mechanism in 802.11 standards. OFDMA allows simultaneous transmissions to/from multiple STAs in the downlink and uplink, and provides more transmission opportunities through frequency-domain enhancements. The introduction of OFDMA improves the multiuser access capability and transmission efficiency in dense deployments.

To simplify OFDMA-based scheduling, 802.11ax divides existing 20, 40, 80, and 160 MHz channel bandwidths into different RUs, and data of each STA is transmitted on only one RU. 802.11ax defines seven types of RUs:

- 26-tone RU

- 52-tone RU

- 106-tone RU

- 242-tone RU

- 484-tone RU, which applies only to 40, 80, and 160 MHz

- 996-tone RU, which applies only to 80 and 160 MHz

- 2 × 996-tone RU, which applies only to 160 MHz

Figure 2.51 illustrates RU division in 20, 40, and 80 MHz for OFDMA operation.

In addition to the introduction of OFDMA, 802.11ax enhances downlink MU-MIMO and introduces uplink MU-MIMO. Figure 2.52 illustrates an uplink MU-MIMO example.

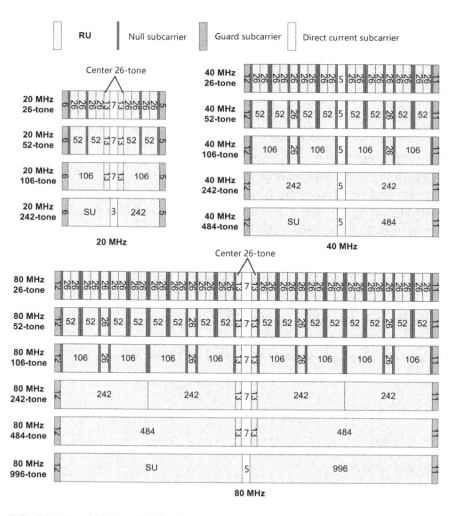

FIGURE 2.51   802.11ax RU division.

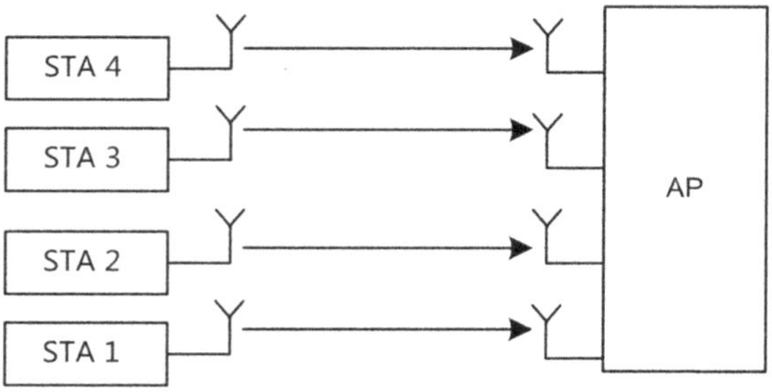

FIGURE 2.52   Uplink MU-MIMO example.

Regarding downlink MU-MIMO, 802.11ax enables an AP to support MU-MIMO transmissions on an entire channel or an RU, and allows for simultaneous scheduling of different RUs for MU-MIMO operation. Whereas 802.11ac supports four users in one MU-MIMO transmission, 802.11ax supports up to eight.

802.11ax introduces uplink MU-MIMO to balance uplink and downlink throughput. Uplink MU-MIMO, similar to downlink MU-MIMO, also enables MU-MIMO transmissions by an AP on an entire channel or an RU, and allows for simultaneous scheduling of different RUs for MU-MIMO transmissions, with up to eight users supported in one transmission.

It should be noted that the introduction of uplink and downlink OFDMA facilitates respective acknowledgments in response to downlink MU-MIMO and uplink MU-MIMO, and reduces acknowledgment transmission overheads. This feature is yet another advantage of MU-MIMO in 802.11ax.

2.4.4.2 Multiuser Transmission

The multiuser transmission feature distinguishes 802.11ax from earlier Wi-Fi standards. Whereas 802.11ac supports only downlink MU-MIMO, 802.11ax supports both uplink and downlink OFDMA transmissions as well as uplink and downlink MU-MIMO. To support efficient multiuser transmissions, 802.11ax optimizes the buffer status report (BSR) and enhances the mechanisms for aggregate media access control protocol data unit (A-MPDU) fragment transmissions and channel measurement. 802.11ax also proposes channel protection and multiuser EDCA for multiuser transmissions, as well as OFDMA-based random access.

To efficiently obtain user transmission requirements, 802.11ax proposes a trigger frame-based NDP feedback mechanism. As such, an AP simultaneously collects the transmission requirements of multiple STAs, upon which efficient multiuser scheduling is possible.

1. Downlink multiuser transmissions

    A downlink multiuser transmission procedure, as illustrated in Figure 2.53, is initiated by an AP. In most cases, an AP sends downlink multiuser data frames to STAs using the HE MU PPDU format, and once these data frames are received, STAs reply with ACK frames as specified.

    In downlink multiuser transmissions, an AP sends data frames to multiple STAs. An STA is allocated only one RU to transmit data, and this RU is also used for transmitting HE-STF (for AGC) and HE-LTF (for channel estimation). When a bandwidth provides multiple 20 MHz channels, the RU size may be less than or greater than 20 MHz. Legacy preambles (L-STF, L-LTF, and L-SIG), RL-SIG, and HE-SIG-A are transmitted on 20 MHz channels separate to the primary 20 MHz channel. This repetition mechanism ensures that STAs do not experience interference with burst signals on each channel within the bandwidth in downlink OFDMA operation. Figure 2.54 illustrates the PPDU format for downlink multiuser transmissions in 80 MHz.

    STAs send ACK frames in the following three scenarios:

    • STAs send Block Acknowledgment (BA) frames using OFDMA in response to trigger information sent from an AP. The trigger

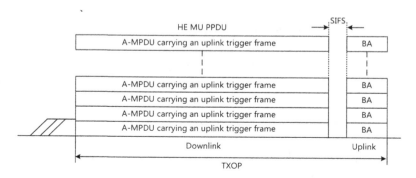

FIGURE 2.53 Downlink multiuser transmission procedure.

	L-STF	L-LTF	L-SIG	RL-SIG	HE-SIG-A	HE-SIG-B1	HE-STF #1	HE-LTF #1	Data #1
							HE-STF #2	HE-LTF #2	Data #2
	L-STF	L-LTF	L-SIG	RL-SIG	HE-SIG-A	HE-SIG-B2	HE-STF #3	HE-LTF #3	Data #3
80 MHz	L-STF	L-LTF	L-SIG	RL-SIG	HE-SIG-A	HE-SIG-B1	HE-STF #4	HE-LTF #4	Data #4
	L-STF	L-LTF	L-SIG	RL-SIG	HE-SIG-A	HE-SIG-B2	HE-STF #5	HE-LTF #5	Data #5

FIGURE 2.54   PPDU format for downlink multiuser transmissions.

information is either carried in the MAC header of a downlink data frame or a trigger frame aggregated with a data frame, as illustrated in Figure 2.53.

- An AP sends a Block Acknowledgment Request (BAR) to trigger ACK frames. The acknowledgment mode is signaled in the MAC header of a downlink data frame.

- An AP sends a Multiuser Block Acknowledge Request (MU-BAR) frame to request multiple STAs to transmit ACK frames.

2. Uplink multiuser transmissions

An uplink multiuser transmission procedure, as illustrated in Figure 2.55, is initiated after a trigger frame, carrying STA identifiers and resource allocation information, is sent from an AP. After receiving the trigger frame, STAs send uplink data frames on allocated RUs using the HE TB PPDU format and receive ACK frames sent from the AP after an SIFS.

In trigger frame-based uplink multiuser transmissions, STAs only send their data portions (HE-STF, HE-LTF, and Data fields) on the RUs allocated to them. Assume that an allocated RU is

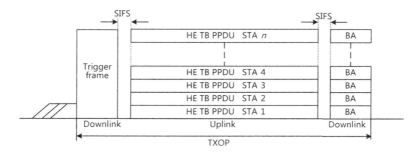

FIGURE 2.55   Uplink multiuser transmission procedure.

located within a 20 MHz channel. The data portions of STA 2 (HE-STF#2, HE-LTF#2, and Data#2) are sent on one allocated RU, and legacy preambles, RL-SIG, and HE-SIG-A of STA 2 are sent on the entire 20 MHz channel within which the allocated RU is located, as illustrated in Figure 2.56. It should be noted that the channel for sending legacy preambles, RL-SIG, and HE-SIG-A is in a minimum unit of 20 MHz. Such information is only sent on the 20 MHz channel within which the target RU is located, and not on any other 20 MHz channels. This mechanism prevents an STA's legacy preambles, RL-SIG, and HE-SIG-A from occupying all the 20 MHz channels available in the bandwidth, enabling STAs in other BSSs to compete for idle 20 MHz channels and improving overall spectrum efficiency. In Figure 2.56, frequency-domain resources used by other STAs for transmissions are presented within broken lines.

An AP can send ACK frames in multiple modes. These include simultaneously sending a BA frame to multiple STAs using the HE MU PPDU format, or broadcasting a Multistation Block Acknowledgment (MBA) frame to signal acknowledgment information destined for all STAs.

802.11ax defines various types of trigger frames, such as those used to trigger uplink multiuser transmissions, known as Basic Trigger Frame (BTF), and those used to trigger control frames from multiple users, such as Multiuser Request to Send (MU-RTS) trigger frames, MU-BAR trigger frames, Beamforming Report Poll (BFRP) trigger frames, BSR poll trigger frames, and NDP Feedback Report Poll (NFRP) trigger frames. The functions of these trigger frames will be detailed in subsequent descriptions.

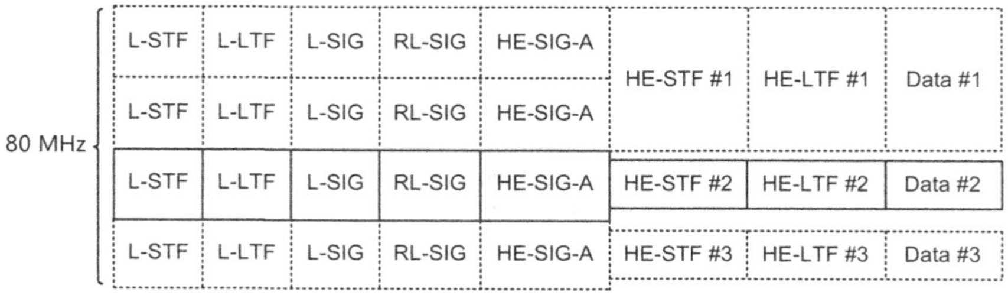

80 MHz	L-STF	L-LTF	L-SIG	RL-SIG	HE-SIG-A			
	L-STF	L-LTF	L-SIG	RL-SIG	HE-SIG-A	HE-STF #1	HE-LTF #1	Data #1
	L-STF	L-LTF	L-SIG	RL-SIG	HE-SIG-A	HE-STF #2	HE-LTF #2	Data #2
	L-STF	L-LTF	L-SIG	RL-SIG	HE-SIG-A	HE-STF #3	HE-LTF #3	Data #3

FIGURE 2.56   PPDU format for uplink multiuser transmissions.

2.4.4.3 BSR

In trigger frame-based uplink multiuser transmissions, an AP signals a certain amount of allocated resources in a trigger frame for STAs to transmit uplink data. The amount of allocated resources is determined by a BSR, which indicates the traffic volume of an STA. Standards earlier than 802.11ax define a BSR reporting mechanism that enables STAs to carry their queue length information in a QoS control subfield for BSR reporting. However, with this mechanism, STAs can report a queue length of only one service type at a time. As many service types are available, STAs send many frames, and this leads to relatively low efficiency.

To enable an AP to efficiently obtain the traffic volume in the STA's buffer, 802.11ax defines a novel BSR reporting mechanism that enables a BSR control subfield to signal BSR information. Figure 2.57 illustrates the structure of a BSR control subfield. Table 2.6 explains the parameters involved.

Scaling Factor is introduced to address the limited number of bits (both Queue Size High and Queue Size All are only eight bits long). If

FIGURE 2.57   BSR control subfield.

TABLE 2.6   Descriptions of Parameters in a BSR Control Subfield

Parameter	Number of Bits	Description
ACI Bitmap	4	Indicates the ACs for which data stored in the STA's buffer is intended. Each bit indicates the presence of a service intended for an AC
Delta TID	2	Indicates the total number of traffic IDs (TIDs) in the buffer
ACI High	2	Indicates the STA-perceived AC with the highest priority and
Queue Size High	8	the corresponding queue length. After obtaining this information, an AP can preferentially schedule the AC with the highest priority in the trigger frame to improve the STA's QoS
Queue Size All	8	Indicates the total traffic volume of data packets in all queues in the STA's buffer, based on which an AP calculates the total number of resources required by the STA
Scaling Factor	2	Indicates four types of queue size units (16, 256, 2048, and 32,768 bytes)

a small static unit is used, it is difficult to indicate a large traffic volume. Conversely, if a large static unit is used, it is difficult to precisely represent a small or medium-sized traffic volume. The use of Scaling Factor delivers high precision when the queue size is small or medium, and expresses a large scope when the queue size is large.

In addition to an efficient BSR control subfield, an AP in compliance with 802.11ax can send a BSR poll trigger frame to multiple STAs in order to obtain BSR information, thereby improving BSR acquisition efficiency.

2.4.4.4 Fragment Transmission and A-MPDU Enhancement

Before fragmentation and aggregation are examined, some concepts related to both the MAC and physical layers in 802.11 are first introduced. As illustrated in Figure 2.58, once the MAC layer receives a MAC service data unit (MSDU) from the logical link control (LLC) layer, it encapsulates the MSDU into a MAC protocol data unit (MPDU), which includes a MAC header and a Frame Check Sequence (FCS) in the tail. The MAC management sublayer of the MAC layer encapsulates management packets into MAC management protocol data units (MMPDUs). The physical

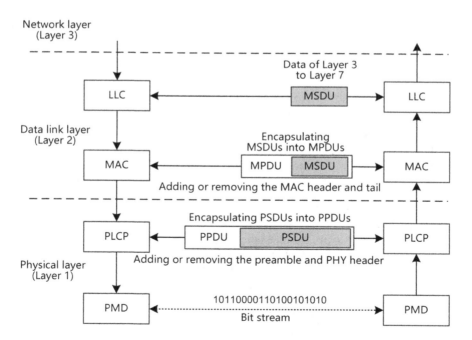

FIGURE 2.58   Basic process of sending and receiving WLAN packets.

layer consists of the physical layer convergence procedure (PLCP) layer and the physical medium dependent (PMD) layer.

After being transmitted from the MAC layer to the PLCP layer, the MPDU receives a new identity, and becomes known as a PLCP service data unit (PSDU). In reality, both MPDU and PSDU refer to the same thing and differ in name only. After receiving a PSDU, the PLCP layer adds a preamble and a PHY header to this frame to form a PPDU, and then transmits the PPDU to the PMD layer. The PPDU is then modulated into a stream of bits (0/1) for transmission, depending on different algorithms.

With fragmentation, an MSDU or an MMPDU at the MAC layer is divided into multiple fragments of the same size (except for the last fragment). Each fragment is called an MSDU or MMPDU fragment. When a fragment fails to be received, only that fragment is retransmitted, not the entire MSDU or MMPDU. This mechanism improves network robustness and throughput.

To support an increased data transmission rate, 802.11n introduces aggregate MAC service data units (A-MSDUs). Many fixed MAC overheads exist in 802.11 standards, especially acknowledgment information between two frames. 802.11 stipulates that each time a unicast data frame is received, an ACK frame must be immediately returned. At the highest transmission rate, these extra overheads are even larger than the entire data frame being transmitted. With A-MPDU, an MSDU or an A-MSDU is encapsulated into an MPDU, and multiple MPDUs are aggregated into a physical layer packet. N MPDUs can be simultaneously transmitted through only one channel contention or backoff. This mechanism does not cause channel resource consumption that can arise from separate transmissions of the remaining $N-1$ MPDUs. The aggregated MPDUs are distinguished by MPDU delimiters, and an A-MPDU contains a maximum of 64 MSDUs. Figure 2.59 illustrates how this mechanism works.

After receiving an A-MPDU, the receiver processes each MPDU contained within and sends an ACK frame in response to each one. To reduce the number of ACK frames in this case, the BA mechanism is employed, which uses one ACK frame to respond to multiple MPDUs. Figure 2.60 illustrates how this mechanism works.

Aggregation and fragmentation are mutually exclusive.

OFDMA in 802.11ax enables multiple STAs to transmit data over different subchannels, as illustrated in Figure 2.61. Individual STAs transmit A-MPDUs over their subchannels. One MSDU or one A-MSDU, not

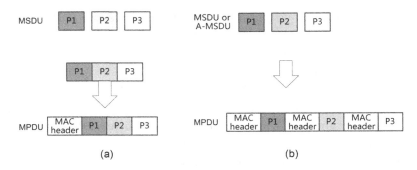

FIGURE 2.59   (a) A-MSDU and (b) A-MPDU.

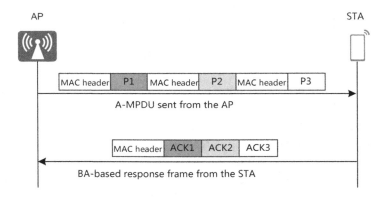

FIGURE 2.60   A-MPDU BA.

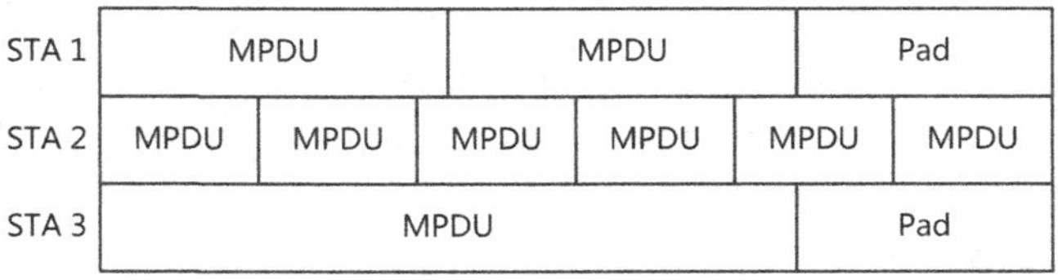

STA 1	MPDU		MPDU		Pad	
STA 2	MPDU	MPDU	MPDU	MPDU	MPDU	MPDU
STA 3	MPDU				Pad	

FIGURE 2.61   A-MPDU in OFDMA.

one MSDU fragment, is encapsulated into one MPDU in an A-MPDU. To align data transmissions in the time domain, STAs append padding bits.

To improve transmission efficiency, 802.11ax proposes dynamic fragmentation. This mechanism enables MSDU, A-MSDU, and MMPDU fragments of varying sizes, and allows for aggregation of fragments in A-MPDUs, leading to improved transmission efficiency in OFDMA operation. In addition, one MSDU or A-MSDU fragment can be transmitted over resources intended for padding bits, further improving transmission efficiency.

Under 802.11ax fragmentation, the following three levels apply:

- Level 1: A single MPDU (not an A-MPDU) carries one MSDU, A-MSDU, or MMPDU fragment. A recipient responds with an ACK frame.

- Level 2: An A-MPDU carries multiple fragments. Under the presence of multiple MSDUs or A-MSDUs, an A-MPDU carries one fragment per MSDU or A-MSDU, and only one MMPDU fragment. A recipient responds with a BA frame. Each bit in the bitmap of the BA frame corresponds to one MSDU (or one MSDU fragment) or one A-MSDU (or one A-MSDU fragment).

- Level 3: An A-MPDU carries multiple fragments. Under the presence of multiple MSDUs or A-MSDUs, an A-MPDU carries one to four fragments per MSDU or A-MSDU, and only one MMPDU fragment. Recipients respond with two types of frames. When the Fragment Number subfield is not 0 in at least one received MPDU (0 indicates the MPDU carries an entire MSDU or the first MSDU fragment, or an entire A-MSDU or the first A-MSDU fragment), recipients respond with a fragment BA frame, and every four bits in the bitmap of the fragment BA frame correspond to one MSDU (or four MSDU fragments) or one A-MSDU (or four A-MSDU fragments). When the Fragment Number subfield is 0 in all the received MPDUs, recipients respond with a BA frame, and each bit in the bitmap of the BA frame corresponds to one MSDU (or the first MSDU fragment) or one A-MSDU (or the first A-MSDU fragment). Fragmentation level 3 improves flexibility and OFDMA transmission efficiency.

To match the improved transmission rates, 802.11ax enhances A-MPDU design by extending lengths and types of services that can be aggregated. On the one hand, 802.11ax improves transmission efficiency by increasing the maximum number of MPDUs that can be aggregated in each A-MPDU from 64 to 256. This allows STAs to send more data packets at a time. On the other hand, 802.11ax improves transmission efficiency by introducing multi-TID aggregation enabling STAs to aggregate data packets with multiple TIDs into one A-MPDU. In standards earlier than 802.11ax, only data packets with a specific type of TID can be aggregated into an A-MPDU. This mechanism has two shortcomings: low efficiency and resource waste. Specifically, if an STA runs multiple types of services, all data packets need to be aggregated into multiple short A-MPDUs for transmission. In uplink multiuser transmissions, padding bits are appended to data frames if the data frame length varies among STAs, in order to ensure the same data frame length for all STAs.

2.4.4.5 Channel Measurement

802.11ac defines a channel sounding protocol, also known as the VHT Sounding Protocol, to allow transmitters to obtain CSI from receivers through packet exchange between them, thereby enabling improved beamforming.

The beamformer, according to 802.11ac, first sends an NDP announcement to initialize beamforming. That is, this NDP announcement identifies STAs that are required to send CSI feedback. Then, the NDP announcement is followed by an NDP intended for detection, channel estimation, and time synchronization, after an SIFS. However, if an NDP announcement contains multiple STA Info fields, the NDP announcement must be broadcast. In this case, the first predetermined beamformee will respond with a VHT compressed beamforming frame. With VHT compressed beamforming, the beamformee calculates the beamforming weight based on which compression is performed to reduce system overheads. Then, the beamformer adjusts transmit parameters and performs beamforming after receiving the VHT compressed beamforming frame. If the NDP announcement is intended for multiple predetermined recipients, the remaining predetermined recipients need to wait before response upon polling. Figure 2.62 illustrates the 802.11ac channel measurement mechanism.

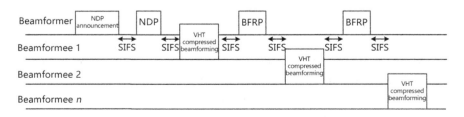

FIGURE 2.62   802.11ac channel measurement mechanism.

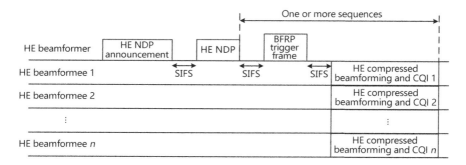

FIGURE 2.63 Channel measurement in 802.11ax.

OFDMA in 802.11ax raises new channel measurement requirements. In addition to channel beamforming, the transmitter also allocates resources for OFDMA operation. As the performance of channel measurement directly affects system throughput, 802.11ax offers improved channel measurement compared to 802.11ac, as illustrated in Figure 2.63. An AP typically sends a BFRP trigger frame to multiple STAs to trigger simultaneous channel measurement result feedback from multiple STAs.

To support uplink and downlink OFDMA transmission, an AP allocates proper RUs to STAs for them to transmit and receive data. In most cases, the STA's channel quality indicator (CQI) is considered during allocation, that is, an AP allocates corresponding RUs to STAs with a desirable CQI. What's more, in 802.11ax, CQI feedback from STAs is supported, offering improved scheduling and increased resource utilization.

2.4.4.6 Channel Protection

On a WLAN, CCA is performed to achieve physical carrier sensing in the channel contention process to avoid signal collision during transmission and, more simply put, to prevent the transmission initiated by an STA from interfering with the ongoing data transmissions of other STAs. However, under the presence of hidden STAs, physical carrier sensing cannot ensure that transmissions initiated by an STA do not interfere with ongoing reception on another STA. To address this challenge, early 802.11 standards introduced virtual carrier sensing, where an NAV is set to ensure that both the transmitter and receiver of a transmission are not interfered by other STAs. In virtual carrier sensing, the transmitter and the receiver are both protected from interference during a period of time through RTS/CTS frame exchange before a transmission starts, as illustrated in Figure 2.64.

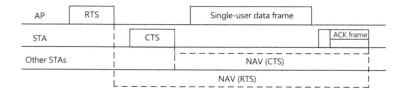

FIGURE 2.64   RTS/CTS mechanism.

In 802.11ax, multiuser transmission is widely applied. If an AP implements channel protection through sequential RTS/CTS frame exchange with each STA before a multiuser transmission starts, a large overhead arises. Therefore, 802.11ax introduces the MU-RTS mechanism to enable an AP to implement simultaneous channel protection through one MU-RTS/CTS frame exchange with multiple STAs, significantly improving system efficiency and reducing large overheads.

With the MU-RTS/CTS mechanism, an AP sends an MU-RTS trigger frame to enable multiple STAs to simultaneously send CTS frames. As the AP can exchange data with multiple STAs on the same 20 MHz channel, CTS frames also need to be sent on this channel. To ensure that the AP correctly receives CTS frames from multiple STAs, the content of CTS frames from all these STAs must be identical. In addition, all the CTS frames must be transmitted in the same manner, such as using the same MCS and scrambling scheme. To meet this requirement, an AP can signal a CTS frame transmission manner in the MU-RTS frame, with transmit parameters of all CTS frames set to the same value and the scrambling code set to the code used by the MU-RTS trigger frame.

When using the MU-RTS mechanism, channel protection can be implemented through MU-RTS/CTS frame exchange in both uplink and downlink multiuser transmissions. A downlink multiuser transmission is used as an example in Figure 2.65. In this example, an AP sends an

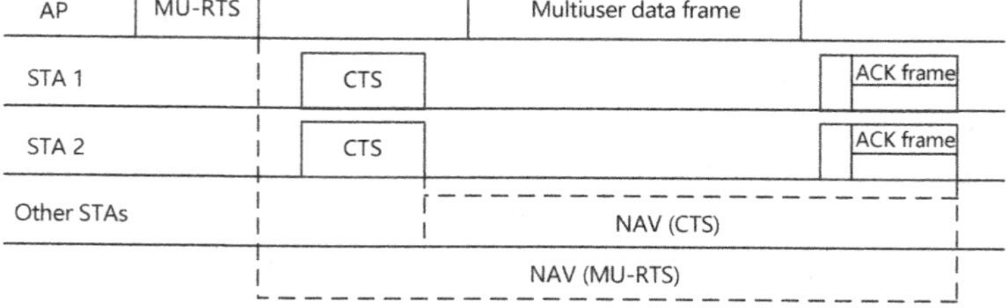

FIGURE 2.65   MU-RTS/CTS mechanism.

MU-RTS frame to STA 1 and STA 2, and the other STAs set an NAV after receiving the MU-RTS frame and the CTS frame to ensure that channel access is not performed before the multiuser transmission is completed.

2.4.4.7 MU EDCA

Unlike 802.11ac and earlier standards, 802.11ax-compliant STAs can transmit uplink data in two schemes: EDCA contention-based transmission and trigger frame-based transmission. By transmitting data using these two transmissions, 802.11ax-compliant STAs are afforded more channel access opportunities compared to STAs that only comply with earlier standards. To ensure STAs are treated fairly regardless of their compliant standard, 802.11ax introduces the MU EDCA mechanism to ensure that an STA that complies with earlier standards is not severely disadvantaged during channel contention. In addition, as trigger frame-based uplink transmission is more efficient than contention-based uplink transmission, MU EDCA is used to increase channel access opportunities for APs to transmit trigger frames to schedule uplink transmissions.

MU EDCA also increases the access priority in legacy EDCA after an STA is triggered by an AP to send data. More specifically, an STA that supports MU EDCA uses a different EDCA parameter set, also known as an MU EDCA parameter set, to define a longer time period of contention waiting and backoff. As such, the legacy EDCA parameter set has a higher priority than an MU EDCA parameter set.

An AP signals two EDCA parameter sets in a Beacon frame or an association response frame. One is a legacy EDCA parameter set, and the other is an MU EDCA parameter set. As mentioned above, the MU EDCA parameter set includes a larger arbitration interframe spacing number (AIFSN), a minimum contention window (CWmin), and a maximum contention window (CWmax). Then, when an STA is not triggered to send data, it performs EDCA contention with a legacy EDCA parameter set. The STA, after being triggered by a trigger frame from an AP to send uplink data, performs channel contention with the MU EDCA parameter set within a specific period carried in the MU EDCA parameter set. However, if the STA does not receive a new trigger frame within the specific period and completes the uplink data transmission, the STA performs channel contention with the EDCA parameter set (legacy EDCA contention). Figure 2.66 illustrates the MU EDCA working principle.

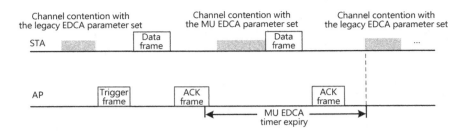

FIGURE 2.66   MU EDCA working principle.

2.4.4.8 OFDMA-Based Random Access

In 802.11ax, with uplink OFDMA introduced, an AP sends a trigger frame to schedule multiple STAs on different RUs to enable the STAs to simultaneously transmit uplink data. However, as uplink OFDMA-based scheduling is not effective in some scenarios, the uplink OFDMA-based random access (UORA) mechanism is used instead. These scenarios are described in more detail below:

- Scenario 1: If an STA's uplink service is aperiodic, or an AP exits the sleep state and does not know which STAs need to send uplink data, the AP employs UORA to enable random access of all the STAs that meet uplink access conditions. UORA, with its improved transmission efficiency, is especially suitable for STAs that report BSR or send small packets. For example, when a large overhead arises from single-user preamble transmissions with a small uplink traffic volume, UORA can be employed so that all STAs are allocated a small RU and transmit the same preamble, thereby reducing overheads.

- Scenario 2: In typical application scenarios, as the transmit power of an AP is greater than that of an STA, the coverage area of an AP is also greater than that of an STA. In addition, when an STA is far away from an AP, although the STA can receive Beacon frames from the AP, the uplink packets sent from the STA cannot reach the AP, thereby causing association failure between the STA and the AP. With UORA, however, an STA can aggregate power on an RU with a bandwidth less than 20 MHz to increase the transmit power. By doing this, the AP can receive uplink packets from the STA. Then, after the STA is associated with the AP, the STA can communicate with the AP through either uplink OFDMA-based scheduling or UORA.

- Scenario 3: 802.11ax defines several RU allocation methods where each STA is allocated only one RU. In most cases, an AP does not allocate all RUs, resulting in a waste of frequency-domain resources. For example, a 20 MHz channel may be divided into four 52-tone RUs and one 26-tone RU. If three STAs have uplink data to transmit, the AP can allocate one 52-tone RU to each STA, while the remaining 52-tone RU and 26-tone RU remain unallocated. In this case, the unallocated RUs can be used in random access through UORA, ensuring channel resources are fully utilized.

As UORA is only implemented at the MAC layer, it does not entail any changes at the physical layer. When an AP signals RU allocation for random access in a trigger frame, the trigger frame includes one or more user information fields (UIFs). AID12, a subfield in a UIF, usually indicates the twelfth bit of an association ID (AID) of an STA. An AID is the unique ID sent from an AP to an STA to identify their connection when the STA is being associated with the AP. What's more, when the AID12 is set to 0, the RU corresponding to the UIF is allocated to an associated STA. And when the AID12 is set to 2045, the RU corresponding to the UIF is allocated to an unassociated STA. In trigger frame-based scheduling scenarios, each UIF can signal only one RU allocated to an STA. To reduce signaling overheads, UORA allows one UIF to signal allocation of up to 32 consecutive RUs of the same size to multiple STAs. What's more, in one trigger frame, RUs can be allocated for scheduling STAs, while some RUs can also be allocated for UORA. In addition, one trigger frame can signal RUs allocated to both associated STAs and unassociated STAs for UORA. As illustrated in Figure 2.67, the AP allocates RU 1, RU 2, and RU 3 to three associated STAs (STA 1, STA 2, and STA 3) with AIDs 1, 5, and 7, respectively. It also associates RU 4, RU 5, and RU 6 to the associated STAs for UORA, and RU 7, RU 8, and RU 9 to the unassociated STAs for UORA. In HE TB PPDUs that follow the trigger frame after an SIFS, scheduled STA 1, STA 2, and STA 3 separately transmit uplink data on allocated RUs, the associated STA 4 and STA 5 respectively transmit uplink data on RU 4 and RU 6, and the unassociated STA 6 transmits uplink data on RU 9 by

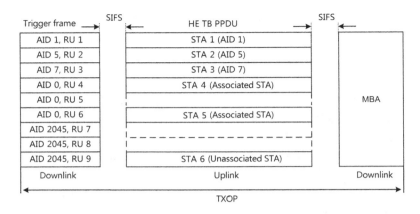

FIGURE 2.67   UORA mechanism.

contention. In Figure 2.67, RU 5, RU 7, and RU 8 are not allocated to any STAs, meaning that these RUs are idle in the HE TB PPDU transmission.

To ensure that UORA is most effective, a UORA backoff mechanism is introduced to STAs, while the legacy EDCA backoff mechanism is also retained. The two backoff mechanisms operate independently.

OFDMA contention window (OCW) and OFDMA backoff (OBO) are used in the UORA backoff mechanism. An AP carries OCW parameters (OCWmin and OCWmax) in the Beacon frame. After receiving the Beacon frame, a UORA-capable STA sets the OCW to OCWmin and randomly selects an integer that ranges from 0 to OCW as the initial value of the OBO counter. If an STA that has data in the buffer receives a trigger frame that signals RUs for random access, the STA subtracts the number of RUs for random access from the value of the OBO counter. For an associated STA, however, it subtracts only the number of RUs for which AID12 is set to 0, and for an unassociated STA, it subtracts only the number of RUs for which AID12 is set to 2045. If the result is 0 or a negative number, the STA randomly selects an RU for access to transmit uplink data. By contrast, if the result is greater than 0, the STA continues to back off and waits for the next trigger frame signaling RUs for random access. If an STA receives an expected response frame (usually an MBA frame) or does not need a response frame after an uplink transmission following UORA, the transmission is considered successful, and the STA initializes the OCW to OCWmin. If the STA does not receive the expected response frame, the STA considers the transmission has failed. In this case, if the

OCW is less than OCWmax, the STA updates the OCW to the result of $2 \times OCW + 1$; however, if the OCW is equal to OCWmax, the STA retains the OCW.

2.4.4.9 NDP Feedback

A significant volume of feedback messages in the WLAN system requires 1-bit information. Examples include if a buffer exists and whether a current channel is available. Furthermore, all information is required to be fed back by using a MAC frame, with the MAC frame needing to include fields such as a frame control field, a receiver address, and an FCS. As a result, overheads are large even if a short frame is used to carry 1-bit information. Moreover, feedback efficiency can be significantly improved, if the physical layer could instead carry the 1-bit information.

NDP feedback provides an efficient feedback mechanism. Firstly, an AP sends an NFRP trigger frame to an associated STA, and information, such as the initial AID, bandwidth, and multiplexing flag, is carried in the NFRP trigger frame. Secondly, after receiving the NFRP trigger frame, the STA determines, based on information carried in the NFRP trigger frame, whether the STA is to be scheduled. Finally, the STA sends the AP an NDP Feedback Report Response frame as an HE TB PPDU.

How does an STA determine if it is scheduled? An STA calculates the number of scheduled STAs (N_STA) according to the bandwidth and multiplexing flag. For example, in a 20 MHz channel, 18 or 36 STAs can be scheduled, and as bandwidth increases, the number of STAs scheduled increases, showing there is a linear relationship. If the AID of an STA is greater than or equal to the start AID, however less than the result of start AID + N_STA, then the STA is scheduled. Each scheduled STA can calculate, according to its sequence in all scheduled STAs, the position where it sends the HE NDP Feedback Report Response frame.

This mechanism has been accepted by the 802.11ax standard. Currently, the standard supports only one NFRP trigger frame type, specifically, the Resource Request (RR). The AP sends an NFRP trigger frame of the RR type to an associated STA. The scheduled STA determines, based on whether it needs to send data, if it is required to return an NDP Feedback Report Response frame. Additionally, the NDP Feedback Report Response frame carries the feedback state, which can be a value of 0 or 1. 0 indicates the RR is between 1 and the RR buffer threshold, with 1 indicating that the RR is greater than the RR buffer threshold. Furthermore, the RR buffer threshold is determined by NDP Feedback Report Parameter Set Element carried in the management frame, sent by the associated AP.

The 802.11ax standard also discusses the extension of the NDP feedback report mechanism to unassociated STAs. However, the extension is still under discussion as numerous companies have expressed doubts related to its complexity and necessity.

2.4.5 Anti-interference and Spatial Reuse

Before spatial reuse was introduced, the WLAN system instead used the CSMA/CA method. The CSMA/CA allows only one link to transmit data at a time within the signal coverage of an STA. This can be done only after the STA obtains a channel through contention. Overall, this method enables all WLAN participants in a collision range to evenly share channels. However, as the number of participants greatly increases, transmission efficiency decreases, with this inverse relationship further becoming significantly stronger when there are a large number of APs with overlapping service areas in a network. As illustrated in Figure 2.68, STA 1 is associated with AP 1. From STA 1's perspective, the BSS of AP 1 is an MYBSS of STA 1. STA 1 contends for a channel with STA 2 in the MYBSS, then exchanges data with AP 1. STA 1 is additionally in the coverage of the overlapping basic service set (OBSS), that is, STA 1 can receive packets

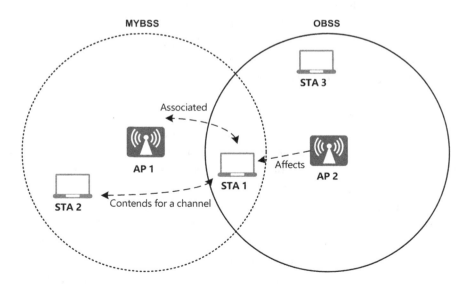

FIGURE 2.68   Reduced transmission efficiency caused by the OBSS.

from AP 2. In this case, communication between STAs and the AP in the OBSS cause STA 1 to back off, and hence STA 1 will be required to wait for a greater time for data transmission. As a result, transmission efficiency reduces. This example depicts a receive frame from the MYBSS referred to as an Intra-BSS frame. It further shows a receive frame from the OBSS referred to as an Inter-BSS frame.

With spatial reuse, after transmission is initiated via the initial link, a different STA in the OBSS determines whether the STA meets a certain condition. This condition specifically stipulates that concurrent transmission cannot affect or must have little impact on the initial link. If the condition is met, concurrent transmission can be initiated.

As a result, spatial reuse improves system throughput. It further assists in determining, as early as possible, if a frame is received from the MYBSS or the OBSS, reducing overall energy consumption. When an STA receives a frame from the MYBSS but the receiver address is not the STA's MAC address, the STA can enter the energy-saving mode during the frame transmission.

Spatial reuse requires an OBSS to perform concurrent transmission. Therefore, Intra-BSS and Inter-BSS frames must be distinguished in the first place. Three following types of criteria are available to distinguish between Intra-BSS and Inter-BSS frames:

1. The receive frame is an Intra-BSS frame if any one of the following three conditions is met:

   • The BSS Color of the receive frame is identical to that of the associated AP.

   • The sender/receiver address of the receive frame is identical to the MAC address of the associated AP.

   • The Partial AID of the receive frame is identical to bits 39–47 in a basic service set identifier (BSSID) of the associated AP.

   If none of these conditions is met, it is an Inter-BSS frame.

2. The AP associated with an STA is a member of a Multiple BSSID set. One physical AP is virtualized into multiple virtual access points (VAPs), and each VAP has an independent BSSID. All BSSIDs of these VAPs compose a Multiple BSSID set. The receive frame is an

Intra-BSS frame if the receiver/sender address of the receive frame matches the MAC address of any member in the Multiple BSSID set, or if the BSSID of the receive frame matches the BSSID of any member in the Multiple BSSID set. Otherwise, the receive frame is an Inter-BSS frame.

3. Some special conditions can additionally be used. For example, if an AP receives a VHT MU PPDU or receives an HE MU PPDU specified in UPLINK_FLAG, the receive frame is an Inter-BSS frame.

Two spatial reuse modes are accepted in the 802.11ax standard: overlapping basic service set–packet detect (OBSS-PD) based spatial reuse and spatial reuse parameter (SRP) based spatial reuse.

2.4.5.1 OBSS-PD-Based Spatial Reuse

With OBSS-PD-based spatial reuse, after an STA receives a frame, if the STA determines it is an Inter-BSS frame and the receive frame strength is less than a specific value (OBSS_PD), the frame is ignored. As a result, the channel state is not assessed as busy and the NAV is not updated based on the receive frame. After a channel backoff is completed based on the OBSS-PD, sending of spatial reuse frames can be initiated. However, when a spatial reuse frame is sent, transmit power is required to be reduced, overall preventing interference to the initial link.

According to different types of OBSS-PD, OBSS-PD-based spatial reuse is further classified into two types:

1. SRG OBSS-PD-based spatial reuse

Spatial reuse group (SRG) is introduced as an OBSS-PD mechanism. SRG is at its most effective when an Intra-BSS signal to interference plus noise ratio (SINR) is far greater than an Inter-BSS SINR. Therefore, an AP can add an OBSS that meets the condition to an SRG set. In this case, the OBSS_PD_min offset has been introduced to make the OBSS_PD_min greater than −82 dBm. As a result, if the receive frame strength is greater than −82 dBm, however less than the OBSS_PD_min, the STA is able to send data without power backoff. The STA needs to send data with power backoff only if the receive frame strength exceeds the OBSS_PD_min, however less than the OBSS_PD.

2. Non-SRG OBSS-PD-based spatial reuse

In such spatial reuse, the OBSS_PD is a value between the OBSS_PD_min and OBSS_PD_max values. The OBSS_PD_max is the sum of −82 dBm and the OBSS_PD_max offset delivered by an AP to an STA. This value prevents OBSS-PD-based spatial reuse from being enabled when the receive frame strength is greater than −62 dBm. The OBSS_PD_min is −82 dBm according to the 802.11ax standard. When an STA detects that the receive frame strength is greater than −82 dBm but less than the OBSS_PD, the STA can send data with power backoff.

As described above, when a spatial reuse frame is sent, transmit power backoff is required and the backoff value is equal to the OBSS_PD minus the OBSS_PD_min. Figure 2.69 illustrates the power backoff.

It must be taken into account that transmit power backoff is still required even if the transmission of a frame that triggers OBSS-PD-based spatial reuse during channel backoff has finished, and the CCA result is less than −82 dBm. In addition, during a secondary channel backoff, OBSS-PD-based spatial reuse can be triggered by multiple receive frames. In this case, the minimum transmit power from multiple power backoffs is used as the transmit power after the completion of channel backoff.

After a transmit opportunity (TXOP) is obtained by the method of OBSS-PD-based spatial reuse, transmit power is required to be

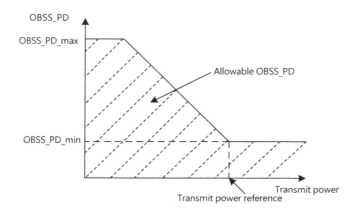

FIGURE 2.69   Power backoff.

less than a backoff power during the entire TXOP period. However, the end time of a TXOP period is allowed to exceed the end time of a receive frame which triggers OBSS-PD-based spatial reuse.

2.4.5.2 SRP-Based Spatial Reuse

Figure 2.70 depicts the principle of SRP-based spatial reuse, where an AP sends a trigger frame, carrying an SRP.

SRP = Transmit power of the trigger frame + Interference strength which can be tolerated by an AP, when the AP receives the HE TB PPDU from an STA

After receiving the trigger frame, STA 3 obtains the trigger frame's transmit power by means of measurement. It is ensured when the frame sent from STA 3 arrives at the AP, the power is less than the interference strength tolerated by the AP, as long as transmit power does not exceed the difference between the SRP and the trigger frame's transmit power.

SRP can be used only if both of the following conditions are met:

- At SIFS time after receiving the trigger frame, STA 3 receives an HE TB PPDU identified as an Inter-BSS PPDU.

- STA 3 has buffered data to be sent, and the to-be-used transmit power is less than the SRP minus a received power level (RPL). The RPL is obtained by measuring an L-STF or L-LTF of the trigger frame.

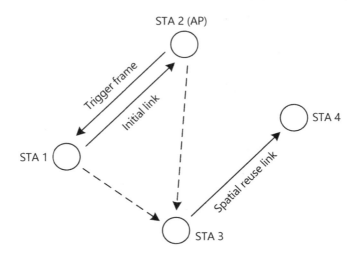

FIGURE 2.70   SRP-based spatial reuse.

After entering the SRP process, STA 3 performs channel backoff, which reuses the backoff counter when SRP is not used. If STA 3 receives another frame during a channel backoff, it needs to be redetermined whether the SRP-based spatial reuse condition is met. If the condition is met, backoff continues. If the condition is not met, the channel is marked as busy. After the value of the backoff counter reaches zero, the end time of the spatial reuse PPDU cannot be later than the earliest end time of multiple SRP_PPDUs, and the transmit power cannot exceed the smallest spatial reuse transmit power.

2.4.5.3 Two NAVs

After uplink multiuser scheduling is introduced to the 802.11ax standard, a new problem occurs for NAV setting. If an STA is scheduled by an AP and the STA's NAV is set by the MYBSS, the STA can respond to the scheduling of the AP. However, when only one NAV exists, the OBSS may set a short NAV for the STA, and then update the NAV to a long NAV set by the MYBSS. In this specific case, when the STA responds to the scheduling of the AP, interference is caused to the OBSS. As a result, the NAV protection set by the OBSS becomes ineffective. Intra-BSS NAV and basic NAV are introduced to resolve this problem.

When an STA receives an Intra-BSS frame whose target address is not the STA's address and the duration of occupying a channel indicated by the frame is greater than the STA's current Intra-BSS NAV, the Intra-BSS NAV is updated. When an STA receives an Inter-BSS frame whose target address is not the STA's address or receives a frame that cannot be distinguished between an Inter-BSS or an Intra-BSS frame, and the duration of occupying a channel indicated by the frame is greater than the STA's current basic NAV, the basic NAV is updated. When at least one NAV is set to a nonzero value, the result of the STA's virtual CS is busy. When two NAVs are both set to a zero value, the result of the STA's CS is idle. When two NAVs are used to determine the channel status, NAVs set for Intra-BSS and Inter-BSS frames will not be overwritten by each other.

Similarly, the industry has additionally discussed the notion that a NAV is set for each OBSS. However, this method has not been adopted by the standard due to its high complexity.

2.4.6 Energy Saving

At initial stages of the proposal of the 802.11 standard, power consumption management has been designed and energy-saving modes are available.

According to this standard, the power of a mobile device is limited, and its power consumption is high if always working. The 802.11ax standard introduces several new energy-saving modes.

2.4.6.1 TWT

TWT was first proposed in the 802.11ah standard. It is designed to save energy for IoT devices, especially devices with low service volume such as smart meters. This allows IoT devices to stay in the sleep state as long as possible, reducing power consumption. After the TWT protocol is established, an STA does not need to receive a Beacon frame but wakes up after a longer period of time. The 802.11ax standard improves the TWT by making rules for STA behavior and implementing channel access control on the premise of meeting energy-saving requirements. TWT is classified into both unicast TWT and broadcast TWT.

1. Unicast TWT

    With unicast TWT, a TWT request STA (a request STA for short) sends a TWT request message to a TWT response STA (a response STA for short), to request to set a wake-up time. After receiving the TWT request message, the response STA sends a TWT response message to the request STA. After the interaction succeeds, a TWT protocol is established between the request STA and the response STA. After the TWT protocol is established, both the request STA and the response STA should keep in an active state in an agreed time period for data transmission and reception; they can enter the sleep state to save energy beyond this period. Generally, an STA (a request STA) sends a TWT protocol setup request to an AP (a response STA). The AP can also initiate a TWT protocol setup request to the STA. After the TWT protocol is established, the negotiated active time is called a TWT service period. One TWT protocol may include multiple TWT service periods each with an equal length, as illustrated in Figure 2.71.

    The start time, duration, and interval of a TWT service period are determined by a TWT parameter set, carried in a TWT request message and a TWT response message. After the TWT parameter set is determined after the TWT negotiation, the STA and the AP can determine the TWT service period.

    Table 2.7 describes three request modes for TWT parameters during a TWT negotiation.

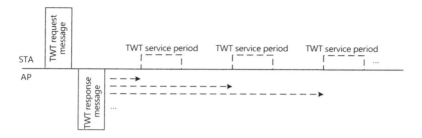

FIGURE 2.71   Working process of unicast TWT.

TABLE 2.7   Different Request Modes and Operations of the Request STA and the Response STA

Request Mode	Request STA	Response STA
Request	Does not specify TWT parameter values	Specifies TWT parameter values
Suggest	Provides suggested values for a TWT parameter set	Can change TWT parameter values based on preferences
Demand	Provides required values, which cannot be changed, for a TWT parameter set	Accepts or rejects TWT parameter values

Table 2.8 describes four response modes for TWT parameters during a TWT negotiation.

After the TWT protocol is established, the request and response STAs communicate only in the TWT service period, disabling channel contention beyond this period. Through this process, the AP can stagger TWT service periods when establishing TWT protocols. This reduces the number of STAs contending for a channel simultaneously, while also managing and controlling the STAs' channel access. Furthermore, during TWT protocol setup, if the

TABLE 2.8   Different Response Modes and Operations of the Response STA

Response Mode	Response STA
Accept	Accepts the request and establishes the TWT protocol
Change	Does not accept the TWT parameter values, and provides new values
Indicate	Does not accept the TWT parameter values, and provides a unique group of values. The TWT protocol can be established only by using the group of TWT parameter values
Reject	Rejects the TWT protocol setup request

Trigger subfield value is 1, the AP sends at least one trigger frame to an STA in each TWT service period to trigger the transmission of an uplink data frame. Since the STA knows in advance that it can perform uplink transmission using the trigger frame, the STA disables its channel contention in the TWT service period and waits for the trigger frame sent by the AP. As a result, the STA can transmit data according to AP scheduling, reducing conflict probability and improving system performance.

2. Broadcast TWT

Unlike unicast TWT, broadcast TWT is capable of batch management, enabling an AP to establish a series of TWT service periods with multiple STAs. During service periods, STAs must maintain an active state to communicate with the AP.

The AP can carry one or more broadcast TWTs in a Beacon frame. Each broadcast TWT has a corresponding identifier, as well as an AP's MAC address. After receiving a Beacon frame, if an STA intends to join the broadcast TWT, the STA can send a broadcast TWT setup request message to the AP. To initiate a request to join a specific broadcast TWT, a broadcast TWT identifier needs to be specified. After joining the broadcast TWT, the STA can be woken up, based on the TWT service period indicated in the TWT parameter set, to communicate with the AP. If the STA supports the broadcast TWT, but does not explicitly join a broadcast TWT, the STA participates by default in the broadcast TWT whose TWT identifier is 0.

The broadcast TWT parameter set specifies the interval and duration of each TWT service period in a similar way to the unicast TWT. In addition, the broadcast TWT parameter set includes the broadcast TWT's lifecycle in a unit of Beacon IFS, which indicates the duration of an established broadcast TWT.

During a broadcast TWT setup, the AP briefly describes the resources intended for allocation in a TWT service period. For instance, whether a trigger frame is sent in the TWT service period, and whether resources used for random access are allocated to the trigger frame. Additionally, the AP can restrict the type of frame sent by an STA in the TWT service period. For example, whether only a control frame and management frame can be sent, or whether an association request frame is allowed to be sent. This restriction enables different types of frames to be sent using various TWTs, thereby enhancing the resource scheduling capability of the AP.

2.4.6.2 OMI

Operating mode indication (OMI) reduces the power consumption of a device in an active state, whereas TWT shortens active time periods, saving power. After an STA is added to a BSS, the STA reports its transmission capability to the AP, including its maximum bandwidth as well as maximum transmit and receive stream quantities. If an STA has sufficient power, or if it is in a power-on state, it uses maximum bandwidth and maximum stream quantities. However, as the STA's power continuously decreases, it can reduce bandwidth and stream quantities using the OMI to save power and enhance battery endurance.

When the STA expects to adjust its receive/transmit parameters using the OMI, the STA can send a data or management frame with an operation mode control subdomain to the AP. This frame carries the receive/transmit parameters that the STA expects to use. When the AP returns the ACK frame, it indicates that the STA's transmit/receive parameters have been successfully modified and new ones can be used in a subsequent TXOP. When the AP sends a radio frame to the STA, the bandwidth and spatial stream quantity higher than those in receive parameters cannot be used. When the AP sends a trigger frame to the STA, the bandwidth and spatial stream quantity of an allocated RU cannot exceed those in transmit parameters. This ensures that even if the STA is in an active state, only a relatively small RF channel and bandwidth can be used, thereby reducing power consumption.

In addition to the restriction on bandwidths and spatial stream quantities, the STA can temporarily disable its uplink multiuser transmission using the OMI. Currently, mainstream multimedia communication devices usually have multiple wireless communication modules (e.g., Wi-Fi and Bluetooth), but concurrent communication on these modules is not supported. This means that if Bluetooth is being used and an AP sends a trigger frame to a Wi-Fi module (i.e., an STA), the STA cannot respond. Therefore, when a device is using Bluetooth, its Wi-Fi module can temporarily disable uplink multiuser transmission by using the OMI to reduce power consumption.

2.4.6.3 20 MHz-Only STAs

Some terminals, such as wearable or IoT devices, have low-power consumption and complexity. These terminals do not have large data volumes and, therefore, do not require large bandwidth. However, contrary to this, these types of terminals must support 80 MHz channel bandwidth according to the 802.11ax standard.

To resolve this issue, two types of STAs are defined in the 802.11ax standard. The first is STAs that must support 80 MHz channel bandwidth, and optionally support 160 MHz or 80 MHz+80 MHz channel bandwidth. The second is STAs that support only 20 MHz channel bandwidth, also referred to as 20 MHz-only STAs.

A 20 MHz-only STA can communicate with an AP via OFDMA over a 40 or 80 MHz channel. However, the 20 MHz channel's subcarrier mapping is different from that of the 40 or 80 MHz channel (including direct current subcarriers, guard subcarriers, and pilot subcarriers). As a result, some data carriers are lost in such scenarios. If the proportion of lost data carriers is not large, the receiver may recover data on the affected carriers through decoding. If the proportion of lost data carriers is too large, performance deteriorates significantly. According to the link simulation result, the 802.11ax standard prohibits an AP from allocating severely affected RUs to serve a 20 MHz-only STA.

By default, a 20 MHz-only STA works on a primary 20 MHz channel, leading to additional problems. If numerous 20 MHz-only STAs can work only on the primary 20 MHz channel, the secondary channels' resources will be wasted. As illustrated in Figure 2.72, a BSS works on 80 MHz channels that are idle and available. However, when an AP communicates with a 20 MHz-only STA, only the primary 20 MHz channel will be utilized, thereby wasting 60 MHz on secondary channels. Under this scenario, a certain mechanism is required to enable a 20 MHz-only STA to work on

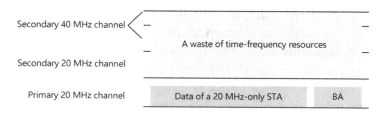

FIGURE 2.72   Resource waste caused by 20 MHz-only STAs.

a secondary channel. In that regard, two ideas were discussed during the standard formulation:

- The first idea involves enabling a 20 MHz-only STA to camp on a secondary channel for a long period of time through AP scheduling. In this method, the STA needs to receive a management frame, such as a Beacon frame, on the secondary channel. However, certain companies are concerned that if a Beacon frame is sent on the secondary channel, coverage for Beacon frames will shrink when power is distributed on a large bandwidth. In addition, 802.11a-compliant STAs may take the secondary channel, which STAs receive Beacon frames from, to be the primary 20 MHz channel.

- The second idea involves temporarily scheduling a secondary channel for the 20 MHz-only STA using the TWT mechanism. The 20 MHz-only STA first operates on the primary 20 MHz channel, and then after receiving the TWT scheduling, it migrates to a specified secondary channel within the TWT service period. After the TWT service period ends, the STA migrates back to the primary 20 MHz channel. This idea has been adopted in the standard because it only requires slight protocol modifications without the need for changing the transmit mode of management frames.

After a secondary channel is scheduled for the 20 MHz-only STA, the EDCA function is disabled. This is because EDCA-based data transmission of the 20 MHz-only STA on the secondary channel cannot be synchronized with that on the primary channel. Consequently, the AP cannot properly receive data from the 20 MHz-only STA. This means that when the 20 MHz-only STA operates on a secondary channel, uplink transmission can only wait for the AP to send the trigger frame for scheduling.

Figure 2.73 illustrates 20 MHz-only STAs operating on secondary channels. The AP uses the TWT mechanism to schedule two 20 MHz channels on the secondary 40 MHz channel for two 20 MHz-only STAs. 80 MHz channel resources can be fully utilized using an HE MU PPDU, while serving 20 MHz-only STAs as follows:

- Downlink data is sent on the primary 20 MHz channel to an STA that supports a bandwidth of 20, 40, or 80 MHz.

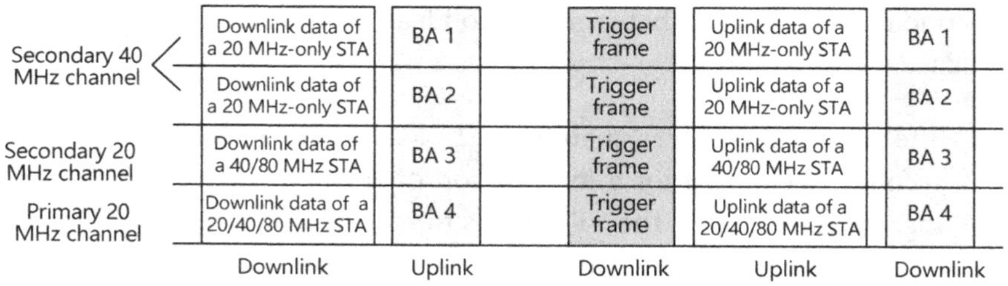

	Downlink data	BA	Trigger frame	Uplink data	BA
Secondary 40 MHz channel	Downlink data of a 20 MHz-only STA	BA 1	Trigger frame	Uplink data of a 20 MHz-only STA	BA 1
	Downlink data of a 20 MHz-only STA	BA 2	Trigger frame	Uplink data of a 20 MHz-only STA	BA 2
Secondary 20 MHz channel	Downlink data of a 40/80 MHz STA	BA 3	Trigger frame	Uplink data of a 40/80 MHz STA	BA 3
Primary 20 MHz channel	Downlink data of a 20/40/80 MHz STA	BA 4	Trigger frame	Uplink data of a 20/40/80 MHz STA	BA 4
	Downlink	Uplink	Downlink	Uplink	Downlink

FIGURE 2.73 Using the TWT for 20 MHz-only STAs.

- Downlink data is sent on the secondary 20 MHz channel to an STA that supports a bandwidth of 40 or 80 MHz.

- Downlink data is sent to two 20 MHz-only STAs on two 20 MHz channels of a secondary 40 MHz channel.

Similarly, the AP can send trigger frames in a non-HT duplicate manner on an 80 MHz channel, and 80 MHz channel resources can be fully utilized while serving 20 MHz-only STAs as follows:

- The primary 20 MHz channel is scheduled for an STA supporting a bandwidth of 20, 40, or 80 MHz to transmit uplink data.

- The secondary 20 MHz channel is scheduled for an STA supporting a bandwidth of 40 or 80 MHz to transmit uplink data.

- Two 20 MHz channels of a secondary 40 MHz channel are scheduled for two 20 MHz-only STAs to transmit uplink data.

2.4.6.4 Others

The uplink and downlink loads of numerous services are asymmetric. For example, after a large Transmission Control Protocol (TCP) packet is transmitted in the downlink, a TCP ACK frame must be sent in the uplink. This process involves the TCP packet and TCP ACK frame being sent by different STAs, and therefore needing to contend for a channel to obtain a TXOP. On the one hand, the number of channel contentions significantly increases, leading to collisions and decreased system efficiency. On the other hand, the obtained TXOP cannot be fully utilized because the TCP ACK frame is very short.

A reverse direction grant (RDG) procedure has been introduced in the 802.11 standard to resolve this issue. Basically, after an STA (referred to

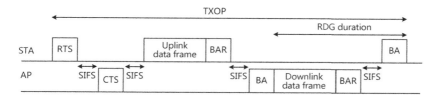

FIGURE 2.74   Basic RDG process.

as an RD initiator) obtains a TXOP through contention, if the STA's data transmission is completed and there is remaining TXOP time, the last frame may indicate that the usage right of the TXOP has been transferred to the specified receive STA (referred to as an RD responder), as illustrated in Figure 2.74. The RD responder can send data to the RD initiator within the remaining TXOP time, ensuring better utilization of channel resources.

After downlink MU-MIMO was introduced into the 802.11ac standard, the RDG usage extended. This means that after an STA serves as an RD initiator and transfers a TXOP to an AP (an RD responder), the AP can send downlink data to multiple STAs through downlink MU-MIMO. However, one of the STAs must be the RD initiator. The method is still being used in the 802.11ax standard.

In addition to downlink MU-MIMO, the 802.11ax standard also supports uplink MU-MIMO, further extending RDG usage. After the STA (the RD initiator) transfers the TXOP to the AP (the RD responder), the AP can schedule multiple STAs via trigger frames to send uplink data through uplink MU-MIMO. The multiple scheduled STAs must include an RD initiator, and the spatial streams allocated to the RD initiator cannot be fewer than the spatial streams of the last frame when the RD initiator performs TXOP transfer. After the AP is allowed to perform uplink MU-MIMO scheduling as an RD responder and the STA obtains the TXOP, the TXOP can be transferred to the AP as early as possible, leading to better utilization of channel resources during the AP's uplink and downlink scheduling.

The following example helps illustrate the RDG mechanism based on uplink MU-MIMO. In Figure 2.75, STA 1 to STA 4 support only one spatial stream, and the AP supports four spatial streams. If STA 1 obtains a TXOP through contention, STA 1 and the AP can mutually communicate using only one spatial stream within the TXOP time period. In

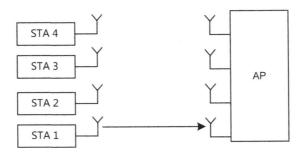

FIGURE 2.75   One STA spatial stream and four AP spatial streams.

this case, although the AP can provide a diversity gain using multiple receive antennas, the throughput cannot be increased using multiple spatial streams.

As illustrated in Figure 2.76, after uplink MU-MIMO is enabled using the RDG, STA 1 to STA 4 support only one spatial stream each. After obtaining a TXOP through contention, STA 1 can use the RDG to transfer TXOP usage rights to the AP. In this case, the AP can simultaneously schedule multiple STAs, including STA 1, through uplink MU-MIMO to use multiple spatial streams for communications. As a result, spatial resources are fully utilized while serving STA 1.

During the standard formulation, the extension from RDG to OFDMA was also discussed. The aim is to schedule an available channel for an STA (an RD initiator) to obtain only one narrowband TXOP. Specifically, if the available channel at the AP (an RD responder) is wider, and the AP performs multiuser scheduling based on OFDMA, the AP's available channel is scheduled for the STA, and secondary channels unavailable to other RD initiators are scheduled for other STAs. As a result, system throughput

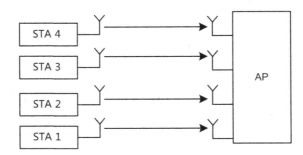

FIGURE 2.76   Uplink MU-MIMO.

improves. However, during the discussion, multiple companies expressed concern related to unfairness and potential interference caused by the OFDMA-based bandwidth extension, and therefore this practice was not accepted in the standard.

2.5 802.11AX PERFORMANCE EVALUATION

During 802.11ax standard formulation, system-level simulation verifies that various technologies with this standard improve WLAN performance. Uplink multiuser transmission is used as an example to verify improvements in 802.11ax compared with 802.11ac.

The simulation uses a typical indoor scenario where one AP is deployed and multiple STAs are positioned at random distances around the AP. Table 2.9 lists specifications for the simulation environment. It is worth noting that 802.11ac uses EDCA contention-based transmission, and 802.11ax uses AP-triggered transmission. The MU EDCA mechanism allows STAs to disable EDCA contention-based transmission, and perform transmission using a trigger frame sent from the AP.

The 802.11 standard defines reference channel model D, which is suitable for indoor work scenarios, such as typical offices, large conference rooms, or open office spaces with partitions.

2.5.1 Impact of Different Transmission Modes on Uplink Throughput

The AP has a fixed coverage radius of 10 m, with the number of STAs increasing from 5 to 15, 50, 100, 200, and 500. Uplink throughput has been measured, as the results in Figure 2.77 indicate.

TABLE 2.9   802.11ax Performance Simulation Environment

Simulation Environment Item	Specifications
Scenario	Indoor
Channel model	Reference model D
Frequency band	5 GHz
Bandwidth	20 MHz
Number of antennas	AP: 4 antennas
	STA: 1 antenna
Transmit power	AP: 20 dBm
	STA: 15 dBm
Traffic model	Full traffic, only uplink data
Packet size	1460 bytes

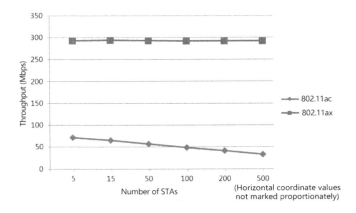

FIGURE 2.77   Impact of different transmission modes on uplink throughput.

- Uplink throughput: The 802.11ax standard supports uplink MU-MIMO, and therefore an AP can receive data from four STAs simultaneously. In this example, five STAs are used, and uplink throughput in 802.11ax is fourfold that in 802.11ac, as illustrated in the figure.

- Transmission efficiency: The 802.11ax standard uses trigger-based transmission. Therefore, the AP schedules STAs to send data, reducing loss caused by STA contention for channel resources. As the number of STAs increases, uplink throughput in 802.11ac decreases; however, this does not occur in 802.11ax. Specifically, as the number of STAs approaches 500, 802.11ax delivers uplink throughput that is over eightfold that in 802.11ac.

2.5.2 Test on Uplink Throughput Improvement Using the AP Scheduling Algorithms

15 STAs were used, and the coverage radius of the AP increased from 10 to 20, 30, 40, and 50 m. The AP uses different algorithms for uplink transmission, and uplink throughput has been measured.

- Random scheduling algorithm

    After contending a channel for use, the AP randomly selects four STAs and triggers data transmission. Figure 2.78 displays the measurement results. When the AP coverage radius is relatively small, uplink throughput in 802.11ax is fourfold that in 802.11ac. However, as the AP coverage radius increases, gain decreases gradually.

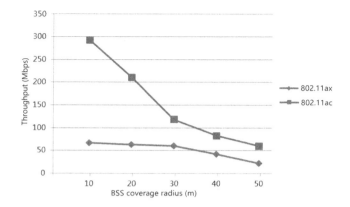

FIGURE 2.78 Impact of the AP scheduling algorithm on uplink throughput (random scheduling algorithm).

- Nearest-STA scheduling algorithm

    Four STAs with optimal channel conditions are always selected for uplink MU-MIMO transmission. Figure 2.79 displays the measurement results. This scheduling algorithm sacrifices fairness for the highest possible uplink throughput.

- *K*-best-STA scheduling algorithm

    Four STAs are randomly selected from *K* STAs with optimal channel conditions for uplink MU-MIMO transmission (*K* equals 8 in this simulation). Figure 2.79 displays the measurement results. This scheduling algorithm prioritizes fairness at the expense of throughput. However, total throughput is about four times higher than that in 802.11ac.

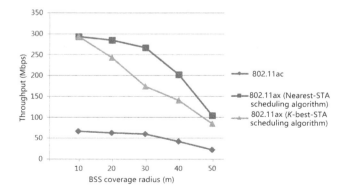

FIGURE 2.79 Impact of the AP scheduling algorithm on uplink throughput.

2.5.3 Test on Throughput Improvement Using Various Technologies

In addition to the preceding simulations, various system-level simulations can be used to verify performance improvement in the 802.11ax standard.

- OFDMA

  In the non-line-of-sight (NLOS) environment, the 802.11ax throughput is 5.9–13 times the 802.11ac throughput. A larger number of STAs indicates a more significant improvement.

- Spatial reuse

  Spatial reuse increases throughput by 30%–60% in an OBSS environment.

2.6 802.11AX STANDARD ACCELERATES INDUSTRY DEVELOPMENT

Enterprise WLANs provide fast and secure wireless Internet access services. Furthermore, with increasingly diverse applications, especially AR/VR, 4K/8K video, and automated guided vehicles (AGVs), wireless networks are transforming into infrastructure for delivering existing and emerging enterprise services. These services require higher bandwidth, lower latency, and larger connectivity capabilities. The next-generation 802.11ax-compliant wireless delivers these capabilities, promoting the application of new technologies and accelerating digital transformation across various industries.

802.11ax (Wi-Fi 6) provides wireless networks with the same level of reliability previously available only on wired networks, thereby enabling a wider range of applications for various industries. For example, a wireless network can be built to replace a wired network to transmit enterprises' production data. Through wireless networks, numerous IoT devices in the manufacturing industry, as well as unmanned robots, can be operated and controlled. While WLAN has been extensively used in scenarios such as enterprise offices, education, healthcare, finance, airports, and manufacturing, 802.11ax will revolutionize these industries.

2.6.1 Enterprise Offices

Mobile offices have gained increasing popularity among enterprises, requiring wireless networks to provide sufficient bandwidth and network reliability. Enterprises will continuously introduce new applications to the existing office mode, enabling efficient cooperation among staff anytime, anywhere.

Enterprise employees use mobile terminals to access their cloud desktops at any location and process services simultaneously. When a problem occurs, employees use a unified communications platform to discuss solutions in a group chat. In addition, desktops can be shared in real time. To achieve the same experience level that previously only wired terminals could provide, real-time video and voice transmission poses high requirements on wireless network latency, less than 20 ms.

Team members use the 4K telepresence conference platform online to hold conference meetings without the need to gather in one room. This is because high-resolution video makes it possible for online and in-person conferences to deliver similar experience levels. Specifically, 4K resolution is 3840 × 2160, which is four times that of 1080p. Therefore, in addition to clearer images, 4K video maintains more natural colors, and this requires a bandwidth of over 50 Mbps as well as latency below 50 ms for video transmission. The more advanced 8K video not only delivers a wide field of view and better image effects but also offers unique image layer and stereoscopic vision effects, which demand bandwidth over 100 Mbps, as listed in Table 2.10.

Wi-Fi 6 improves anti-interference and multiuser service scheduling. Its large network bandwidths help ensure network reliability for new applications. It also improves the efficiency of enterprise offices while driving enterprise office networks to evolve to all-wireless networks.

2.6.2 Education

With the popularization of campus networks, smart mobile terminals and diverse applications are changing the education industry. Specifically, as the number of smart mobile terminals increases, students and teachers expect more multimedia applications, including online lessons, social entertainment, teaching offices, and scientific research management.

TABLE 2.10   4K/8K Video and Wireless Networks Requirements

Resolution	Frame Rate	Color Depth (Bits)	Bandwidth (Mbps)
Entry-level 4K	25/30P	8	>30
Carrier-level 4K	50/60P	10	>50
Excellence-level 4K	100/120P	12	>50
8K	120P	12	>100

Additionally, these users require the same wireless network experience anytime, anywhere, such as in public locations, classrooms, libraries, auditoriums, conference rooms, and dormitories.

When numerous people gather in lecture halls, stadiums, and canteens, there is an exponential increase in data volume on smart terminals. Furthermore, as teachers and students share images and videos in real time using social media applications through campus networks, WLANs must not only provide concurrent access to a large number of users, but also deliver excellent network experience. The Wi-Fi 6 standard improves the concurrent user access capability fourfold while ensuring identical service experience. This completely removes the limitations on traditional Wi-Fi technologies and enables one AP to serve hundreds of STAs.

Teachers utilize VR in classrooms, bringing a brand new learning experience to students. Cloud VR applications based on panoramic video extend experience from one dimension to three dimensions, posing high requirements on wireless networks. If frame freezing occurs, dizziness may occur. Table 2.11 lists VR services and wireless network requirements. The requirements to deliver a comfortable VR experience are network bandwidth and latency of over 75 Mbps and below 15 ms, respectively. Lower latency indicates a better experience. The other types of Cloud VR applications based on computer graphics are games and simulation environments. Unlike panoramic video, these applications involve numerous interactions and require network bandwidth and latency of over 260 Mbps and below 15 ms, respectively. Therefore, creating methods to improve the wireless connection quality of VR terminals has always been an important research topic. The Wi-Fi 6 standard can reduce network latency to less than 20 ms, which lays a solid foundation for offering a comfortable VR experience.

TABLE 2.11   VR Services and Requirements for Wireless Networks

Service	Item	Different Experience Requirements		
		Initial	Comfortable	Premium
Cloud VR video	Bandwidth	>60 Mbps	>75 Mbps	>230 Mbps
	Latency	<20 ms	<15 ms	<8 ms
	Packet loss rate	<0.09%	<0.017%	<0.0017%
Strong-interaction Cloud VR service	Bandwidth	>80 Mbps	>260 Mbps	>1 Gbps
	Latency	<20 ms	<15 ms	<8 ms
	Packet loss rate	<0.01%	<0.01%	<0.001%

In offices and conference rooms, most teachers use laptops on wireless networks for office work, preparing lessons, and communication. Wireless networks can also be used to facilitate collaboration between different campuses and between schools and education departments. Mobile terminals can be used for telepresence conferences anytime, anywhere, with numerous users contributing from various locations.

2.6.3 Healthcare

Patients usually have to wait half to one day before receiving the results of their medical examinations. With Wi-Fi 6, doctors use handheld mobile terminals to access the picture archiving and communication system (PACS) and obtain high definition (HD) check reports such as computerized tomography (CT) and magnetic resonance imaging (MRI) anytime, anywhere, for real-time diagnosis. They can use electronic medical records (EMRs) to query and record patients' conditions in real time, conveniently obtain patients' EMRs during ward rounds, and issue instructions based on patients' conditions.

The popularization of healthcare IoT significantly reduces medical workload. For example, medical personnel had to repeatedly confirm that patients' identities match with the correct medicine to avoid the consequences of prescription errors. With Wi-Fi 6, the mobile infusion management system can process all identity verification. Medical personnel use handheld mobile terminals to scan patients' identity barcodes to obtain relevant information. All processes such as medicine collection, dispensing, and infusion are handled by professional systems, effectively eliminating manual errors. In addition, the healthcare IoT provides multiple solutions, such as medicine management and bed monitoring, to ensure correct patients, medicine, doses, time, and usage.

Wi-Fi 6 also facilitates medical lessons and demonstrations. Traditional surgery demonstrations use wired solutions, which require numerous devices and cables in operating rooms. During surgery, medical personnel may trip over network cables, causing potential injury risks. In the Wi-Fi 6 era, the 4K/VR surgery demonstration system is deployed to synchronize surgery scenarios to each classroom in real time, allowing more medical personnel to participate in observation and learning. In addition, doctors in the operating theater can communicate with remote consultation experts in real time through HD videoconferencing devices.

These new services and applications can be implemented only using Wi-Fi 6. Furthermore, the TWT technology of the Wi-Fi 6 standard can effectively reduce the power consumption of IoT devices and facilitate healthcare IoT.

2.6.4 Finance

Bank service outlets are distributed across various cities. With the emergence of electronic channels such as the mobile Internet, the large-scale competitive advantages accumulated by numerous physical outlets are gradually diminishing, driving these outlets to undergo digital transformation. With the help of wireless networks, banks can use outlets as a gate to mobile interconnection, improving customers' network experience, delivering various personalized services, and building brand images.

New technologies and applications are integrated into outlets to improve the efficiency of service processing. For instance, unattended intelligent outlets have become the new development trend. Before handling business, customers can use a mobile app to query nearby outlets, check queuing status, and make appointments. After arriving at an outlet, customers can join the outlet's wireless network by simply scanning through their mobile phone or accessing the relevant official account. When queuing, customers can browse web pages, view online videos, and enjoy high-speed Internet access services. In addition, customers can interact with the customer reception robot to receive consultation services, query information, or request manual services.

Customers can also use VR devices in an outlet to complete transactions and perform visualized operations on complex investments. Complex financial data is displayed through 3D charts or animations, enabling customers to process and analyze product information intuitively. Customers can use the large HD screen and HD camera of a bank to communicate face to face with consultants or customer service personnel about services and issues.

In addition, outlets can automatically identify VIP customers and provide differentiated services. For example, VIP customers can enjoy more discounts when purchasing products, or with IoT and high-precision positioning technologies, various smart devices and terminals can implement asset management to ensure bank asset security.

In future banking outlets, WLAN will become the main channel for connecting outlets and customers. Furthermore, Wi-Fi 6 will accelerate the application of new technologies and services, making unattended smart outlets a reality.

2.6.5 Airports

With the development and expansion of airports and services, wireless networks have become an integral part of the basic airport network. Requirements for wireless networks are continuously increasing as passengers expect an optimal Internet access experience and various real-time information. From a different perspective, airport operators aim for improved efficiency and service levels, whereas airline staff hope to maintain service stability while improving service accuracy and work efficiency. Finally, airport merchants and advertisers want to increase sales revenue and offer precise product and service pushes.

After arriving at an airport, passengers can use their smart terminals to access a wireless network or use a smart cart to quickly arrive at check-in counters, stores, restaurants, and boarding gates according to a navigation guide. While moving, they receive push information regarding new promotions and discounts from nearby stores. Passengers can browse web pages, make video calls, or view online videos when they are queuing or waiting. In case of flight changes, passengers receive an information push immediately and adjust their plans as prompted. After arriving at a destination airport, passengers move to the baggage area according to the navigation guide, and they can locate their vehicles in the parking lot. Passengers in airports require superior Internet experience and diverse value-added services. In that regard, Wi-Fi 6 is undoubtedly the optimal choice for providing high concurrency and efficiency.

The wireless HD video advertisements, programs, and flight display system provided by airports need 4K/8K to achieve ultra-HD image quality. Therefore, a wireless network must provide a stable and reliable transmission. To this end, high bandwidth and low latency in the Wi-Fi 6 standard ensure real-time image transmission as well as prevent freezing and frame loss caused by unstable signals.

The baggage transportation system uses AGVs to implement unmanned baggage transportation, effectively reducing operating costs and improving transportation efficiency. Although AGVs pose low bandwidth requirements, low transmission latency and high roaming reliability are required within large areas. Generally, AGVs require latency and roaming access time of less than 50 ms, which can easily be achieved on one AGV.

However, if 50 or even 100 AGVs are operating within a 1000 m² area, the wireless network must be well planned to offer latency of 50 ms or even 10 ms. Additionally, the wireless network interface cards for the AGVs must have good compatibility. Furthermore, AGVs must have low-power consumption to ensure long service periods and high transportation efficiency. Wi-Fi 6 features high reliability and energy saving, providing a foundation for unattended baggage transportation systems.

2.6.6 Manufacturing

In the Wi-Fi 6 era, an increasing number of new technologies will be used in manufacturing planning, assisted assembly, detection, and intelligent delivery.

The manufacturing process is complex, requiring factories to provide sufficient space and proper infrastructure. During plant design, VR can be used to perform high-precision digital reconstruction for factories, enabling devices in the plant, such as production lines, pipes, and robots, to be placed in desired positions. VR will facilitate simulation of the actual production process and visualization of each device's running status in a virtual environment. The benefit being that problems will be solved before processes such as factory construction and equipment installation begin.

AR-based collaborative assembly methods can be used in each assembly phase. It facilitates guidance for workers regarding precautions and detailed operations, and sends the surrounding environments to process personnel through interactive means such as voice and marking.

In the manufacturing industry, various detection technologies are becoming increasingly important. These technologies depend on 8K HD videos and images. For example, in the assembly phase, two cameras synchronously obtain product appearance data from two angles. When an error occurs in the product assembly, the background system automatically sends an analysis report to engineering personnel at the station, allowing the problem to be handled on-site. Two to three people were previously required for similar detection tasks, and one measurement takes two hours, resulting in low efficiency. Intelligent detection uses computer vision to check whether the seam width of composite materials meets requirements and whether redundant materials exist, whereas traditional quality inspection methods require on-site personnel. This is inconvenient due to each measurement taking three times longer than intelligent detection.

Unattended operations are the future trend of smart factories. Robots can replace manual operations such as object transportation as well as sorting, and they can process parts on the production line while depending on reliable wireless networks. Wi-Fi 6 networks undoubtedly make smart factories more unmanned.

2.6.7 Next-Generation Wi-Fi

It is foreseeable that with digital industrial transformation, new applications will emerge in quick succession, and service volume will increase sharply, bringing an unprecedented impact on networks. Although the standardization of Wi-Fi 6 has just begun, IEEE 802.11 working group has established the Extremely High Throughput (EHT) Study Group, which will discuss the new 802.11be standard revision based on the Wi-Fi 6 standard. 802.11be aims to increase wireless throughput to 30 Gbps, while further reducing latency and expanding spectrum support to 1–7.25 GHz, which will enable Wi-Fi devices to operate in 2.4, 5, and 6 GHz bands. This will prepare wireless networks for the rapid growth of high-bandwidth and low-latency services. Reliable networks will continue to stimulate the development and popularization of new industry applications and promote industry progress worldwide.

CHAPTER 3

Air Interface Performance and User Experience Improvement

W<small>I-FI HAS ILLUSTRATED STRONG</small> vitality, driven largely by its support for ultra-wide bandwidth. It has seen speeds increase from 11 Mbps in 802.11b to 10 Gbps in the latest standard, contributing heavily to providing users with a significantly improved experience. However, during concurrent usage, the available bandwidth for each user will decrease sharply on a wireless local area network (WLAN), making it difficult to guarantee an optimal experience for each user. This issue is particularly prevalent in office networks where there are a huge number of users per unit area. This chapter will provide an in-depth analysis of the key factors that affect WLAN performance and present some technological solutions for these issues.

3.1 KEY FACTORS AFFECTING WLAN PERFORMANCE

In WLANs, the carrier sense multiple access with collision avoidance (CSMA/CA) mechanism is used to avoid collisions among users contending for radio channels.

On-demand channel access is used in cases where only a small number of users require access, resulting in fewer collisions and sufficient bandwidth for each user. This differs from cellular access technology where each user is allocated with only a limited portion of the channel bandwidth. However, as the number of users increases, CSMA/CA cannot process an excessive number of user access requests, leading to low network performance.

3.1.1 Usable Bandwidth Per User Decreases

An increase in the number of users leads to a shorter time for each user to send data, decreasing the average user throughput and increasing the probability of contention for use of channel resources. Based on the CSMA/CA mechanism, a higher contention probability leads to a higher probability of STAs detecting that a channel is busy, leading to a longer waiting time. This further reduces the chances for users to send packets, resulting in a lower throughput of each user.

3.1.2 User Collision Is Unavoidable

As specified by the CSMA/CA mechanism, STAs detecting that a channel is busy cannot continue sending packets until a certain period of time has passed. With an excessive number of STAs on a network, it is likely that the specified time for two STAs ends at the same time, as illustrated in Figure 3.1. A larger number of STAs lead to a higher probability of this happening.

Following an STA collision, data transmission speed decreases and retransmission is unavoidable. Retransmission requires a new round of contention for channel resources, causing resource waste.

3.1.3 Hidden Nodes Cause Failures in the CSMA/CA Mechanism

Two STAs are considered as hidden nodes if they are at a relatively large distance from each other and unable to receive the probe packets from the other. Both hidden nodes will assume that the channel is clear and begin

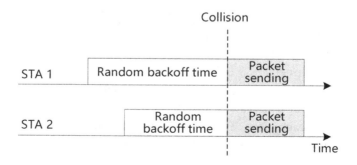

FIGURE 3.1   STA collision.

sending packets to the AP, leading to collision. This means that the AP is unable to correctly receive the packets, and as a result, the CSMA/CA mechanism cannot function properly. This issue can become more severe if the number of STAs on the network increases.

The number of STAs is ever growing, and the network scale and throughput capability delivered by the WLAN technology keep increasing. In the latest edition 802.11ax, known as Wi-Fi 6, orthogonal frequency division multiple access (OFDMA) and trigger-frame-based uplink scheduling have been introduced to address the issues arising from the CSMA/CA mechanism. The new WLAN standard will continue to coexist with earlier standards, with the STAs supporting the new standard to comply with the new scheduling mechanism and those supporting only Wi-Fi 4 and Wi-Fi 5 to comply with the CSMA/CA mechanism. Therefore, it is essential to resolve the multiuser and STA collision issues to improve network performance.

3.2 OPTIMIZATION METHODS FOR WLAN PERFORMANCE

The WLAN's CSMA/CA mechanism is originated from the carrier sense multiple access with collision detection (CSMA/CD) mechanism used in Ethernet. Does Ethernet suffer from the same issues? If so, what are the solutions and can they be applied in WLANs? Before we answer these questions, we will first look at CSMA/CD in more detail.

CSMA/CD works at the Ethernet's Media Access Control (MAC) layer and operates on a "listen before talking and listen while talking" basis. It follows this principle by having a node first listen to the channel (Ethernet cabling), before transmitting, to check whether the channel is clear. If the channel is clear, the node starts to send data while listening to the channel. If the channel becomes busy, the node immediately stops transmission and repeatedly sends an interference packet within a short time, forcing the node in collision to stop transmitting data. Therefore, the CSMA/CD mechanism effectively handles node collision by stopping data transmission to reduce interference.

It is impossible to implement the CSMA/CD mechanism in WLANs because STAs in WLANs work in half-duplex mode and therefore they are unable to listen to channels while transmitting data. To overcome this issue, the CSMA/CA mechanism has been introduced, based on which STAs are instructed to back off, delaying data transmission until a random period of time has passed, avoiding collision.

Therefore, "collision" occurs in the Ethernet too. An excessive number of nodes in a layer-2 network increases the probability of collision, leading to deteriorated network performance, which is similar to WLAN.

The key to resolving this issue in the Ethernet is to narrow down the collision domain. For example, through hardware port isolation on switches, nodes can be confined to several smaller collision domains, significantly reducing the probability of collisions occurring, as illustrated in Figure 3.2.

A similar approach can be implemented in WLANs by deploying multiple APs for traffic diversion. A large WLAN can be divided into multiple small domains to divert STAs to these smaller networks, reducing the number of STAs under each AP to decrease the probability of a collision.

This practice of dividing a large collision domain works only when adjacent APs use different channels, so that STAs in the large WALN can be distributed onto different bands. If STAs are distributed on the same channel, this will make interference worse and increase the number of collisions. WLAN boasts a noticeable advantage in terms of spectrum abundance, particularly the 5 GHz band, which provides an adequate number of channels.

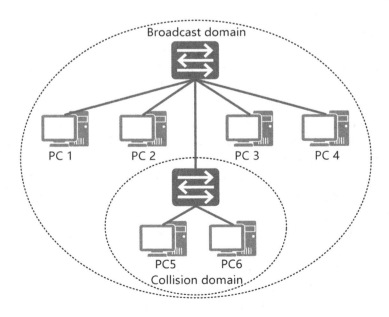

FIGURE 3.2   Collision domain.

TABLE 3.1   WLAN Performance Optimization Measures and Objectives

Performance Optimization Measure		Optimization Objective
Software	Radio calibration	Divide the collision domain, and adjust AP channels and transmit power for maximum capacity usage
	STA roaming and scheduling	Schedule STAs and balance them among APs
	Anti-interference technology	Reduce interference and improve service continuity and communication quality during interference
	Air-interface quality of service (QoS)	Schedule services to improve audio and video service experience
Hardware	Antenna technology	Optimize antennas for improved collision domain division to reduce interference

To fully leverage collision domain division, WLAN has introduced hardware and software optimization (listed in Table 3.1) that facilitates scheduling STAs and services as well as reducing interference.

3.2.1 Radio Calibration

In WLANs, collision domain division correlates with AP channels, and the size of a collision domain is equal to the AP's coverage area, which in turn is related to the AP's transmit power. In view of this relationship, radio calibration aims to choose the appropriate channels and transmit power for APs. For instance, a high transmit power will cause serious co-channel interference and defeat the purpose of deploying multiple APs. A low transmit power will cause insufficient coverage, and STAs in some areas will be unable to access the network.

3.2.2 STA Roaming and Scheduling

In WLAN collision domain division, STAs must be evenly distributed in small collision domains to prevent STAs from overwhelming specific domains. Currently, STAs are self-determining the AP to access and therefore may not choose the most optimal AP. If an STA accesses an AP that is some distance away, the signal will be too weak. In addition, if a large number of STAs access the same AP, the probability of STA collision will subsequently increase.

To overcome these issues, smart roaming, band steering, and load balancing are introduced to WLAN to improve STA scheduling and facilitate preferential connection to optimal APs for STAs based on signal quality, load, and STA capabilities.

3.2.3 Anti-interference Technology

If APs are densely deployed and the channel bandwidth is 40 or 80 MHz, co-channel interference will be inevitable. Using unlicensed Industrial, Scientific, and Medical (ISM) spectrum resources makes WLAN prone to omnipresent external interference from Bluetooth, ZigBee, and other short-range wireless communication devices. Electromagnetic devices, such as microwave ovens, may also cause interference. In this case, the anti-interference technology is used to locate interference sources and to mitigate the impact.

3.2.4 Air-Interface QoS

In each collision domain, voice and video services need to be considered separately, because the CSMA/CA mechanism pursues "fairness." Furthermore, users are very sensitive to voice and video experiences, so these services require a higher transmission priority. Therefore, by introducing enhanced distributed channel access (EDCA) to WLAN, we can classify services into four priorities. The air-interface QoS solutions ensure that parameter settings are flexibly adjusted in response to real-time load change for an improved service experience.

3.2.5 Antenna Technology

Antennas are core components of wireless communication systems. The performance of an antenna is key to improving WLAN coverage, communication quality, and anti-interference effect. Smart antennas that are combined with beamforming technology can mitigate the effect of strong interference. In high-density scenarios, small-angle antennas help scale down collision domains, mitigating the co-channel interference among APs.

3.3 RADIO CALIBRATION

To begin with, let's draw an analogy to help understand radio calibration. As illustrated in Figure 3.3, there are hundreds of people talking, discussing, and holding meetings concurrently in a large room, with each person involved in the same discussion being spread out anywhere between 20 cm and 5 m apart from each other. If the speaker speaks loudly, the people sitting the closest will feel it is too loud, whereas if the speaker speaks quietly, the people sitting far away will not be able to hear clearly.

The easiest solution to this problem is to divide the people in the large room into smaller groups who are talking about the same topic. These

FIGURE 3.3   Effect of speaker volume on the audience.

small groups are then placed into small rooms. This is similar to how radio calibration works. In this section, we will describe how to divide a large "room" into smaller "meeting rooms" by adjusting the location, area, and even wall thickness, in addition to adding or removing "meeting rooms."

3.3.1 Technological Background

Referring to the example mentioned above, multiple APs and a large number of STAs in WLANs are randomly distributed in a domain.

Suppose that these APs work on the same channel and AP 1 has a higher transmit power. As Figure 3.4 illustrates, an STA that is located far away from AP 1 can still receive high-quality downlink signals. However, as the STA has limited transmit power, the uplink signal is unreliable and will consequently suffer delays. As a result, user experience deteriorates significantly.

The solution requires key radio parameter settings to be manually adjusted shortly after a WLAN is deployed. The radio parameters mainly include:

1. Transmit power: An AP's transmit power determines the size of the coverage area of a radio and the level of isolation between cells (which can be compared to "wall thickness" in the example above).

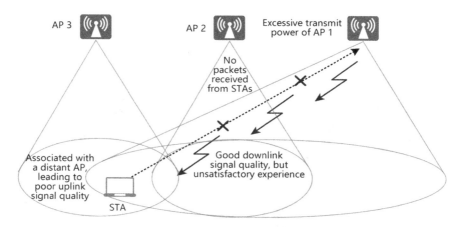

FIGURE 3.4   An AP with high transmit power.

A higher transmit power leads to a higher downlink signal-to-noise ratio (SNR) and makes it easier for the AP to be associated with STAs. However, the transmit power of STAs is limited and is much smaller than APs. Consequently, APs with a high transmit power will be unable to receive signals from the STAs, despite the fact that STAs are able to receive signals from the APs.

2. Band: An AP's band determines the capacity as well as the size of the coverage area of a radio. Two bands are available in WLANs: 2.4 and 5 GHz. The 2.4 GHz band has a relatively smaller number of channels but a significantly lower path loss than the 5 GHz band. This means that the 2.4 GHz band is more prone to co-channel interference in dense networking. Furthermore, the 2.4 GHz band is more likely to suffer from non-Wi-Fi interference caused by cordless phones, Bluetooth devices, microwave ovens, and other devices. Therefore, the 2.4 GHz band provides a far smaller network capacity. For APs that support both 2.4 and 5 GHz bands, radio parameters can be reconfigured so that the 2.4 GHz radio is disabled or works in monitoring mode, or the radio can be configured to work on the 5 GHz band (this requires the APs to support band switching). In ultra-high dense networking, the second 5 GHz radio can be further disabled on the APs with dual 5 GHz radios. Such adjustment will help increase capacity or reduce interference. Referring to the example earlier, this is similar to replacing smaller rooms with a large room.

3. Channel: An AP's channel is also known as a frequency. If two nearby APs work on the same channel, severe contention will occur between them and their STAs. As a result, the throughput will decrease and other channel resources will be wasted. Adjusting channels among APs is similar to "relocating meeting rooms to prevent mutual interference caused by a scramble for projectors, chairs, and other resources."

4. Bandwidth: An AP's bandwidth determines the ultimate throughput, that is, channel capacity. Channel capacity differs vastly between a 20 MHz bandwidth and an 80 MHz bandwidth, so it is essential to assign bandwidth reasonably. This is similar to "merging several medium-sized rooms for a 100-person meeting and rescheduling a three-person meeting to the smallest room."

Radio parameter planning depends on AP locations, interference, and service conditions to determine the transmit power, bands, channels, and bandwidth. Manual configuration is expensive and does not effectively respond to sudden interference. Automatic radio calibration helps automatically detect neighbor relationships between APs, interference on each channel, and load information in a certain period of time. Based on the detection results, the network automatically calculates radio parameters and delivers them to the APs. The following introduces key technologies and explains scenarios that trigger radio calibration.

3.3.2 Automatic Radio Calibration Technology

3.3.2.1 Obtaining Network Status Information

As mentioned above, AP location, interference, and service information are essential for radio parameter planning. Likewise, the information is also needed for automatic radio calibration.

1. RF topology and interference identification

    Radio scanning has been introduced to obtain topology and interference. At the beginning of a radio calibration process, all APs perform scanning for a certain period of time. That is, APs switch to other channels to send probe request frames and receive probe response frames as well as listen to Beacon frames and other 802.11

frames. They further obtain the transmit power on the scanned channels by means of, for example, wireless access controller (WAC) delivery, mutual notification through self-established links, and indication through vendor-custom fields of these frames. The difference between the obtained transmit and receive power on the scanned channels indicates the path loss. The physical distance between the two radios can be derived based on the calculation formula provided by the Institute of Electrical and Electronics Engineers (IEEE) Task Group AC (TGac):

$$d = 10^{\frac{L-20(\lg(f))-p+28}{10D}}$$

where L denotes path loss in dB, f denotes frequency in MHz, D denotes the attenuation factor, d denotes distance in meter, and p denotes the penetration factor.

In semienclosed indoor environments, the following parameter values are used for approximate calculations.

- 2.4 GHz band: attenuation factor $D=2.5$ and penetration factor $p=6$

- 5 GHz band: attenuation factor $D=3$ and penetration factor $p=6$

Based on this formula, distance d can be calculated through frequency f, path loss L, and attenuation factor D. Subsequently, the distance difference between any two APs can be obtained, as can network topology (neighbor) information.

During channel scanning, by listening to 802.11 frames, APs obtain numerous packets, including air-interface packets from their peers within the same WAC and packets from other APs (or wireless routers). The MAC addresses, receive power, and channels of APs that are not within the same WAC and these of the wireless routers will be recorded as external Wi-Fi interference information.

In addition, spectrum scanning (see Section 3.5.1) will also be enabled on APs to collect time-domain signals, which are subsequently subject to fast Fourier transformation (FFT) operations and matched against the spectrum template to recognize non-Wi-Fi interference. The results will also be stored with the receive power and channel information.

Each AP periodically sends the WAC the information it has col-lected about all neighbors, Wi-Fi interference, and non-Wi-Fi inter-ference. After performing processes such as filtering, the WAC creates matrices for the network topology, Wi-Fi interference, and non-Wi-Fi interference, as illustrated in Figure 3.5.

2. RF load statistics

RF load statistics of a radio can be collected with ease. This is because APs record and send the wired and wireless load statistics generated within a period of time, and also report the number of STAs within the period to the WAC. Based on this information,

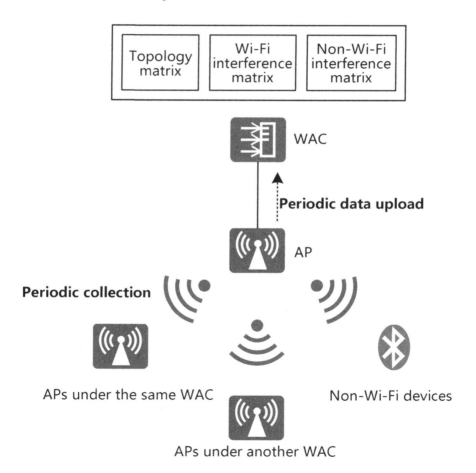

FIGURE 3.5 Principle of RF topology and interference identification.

the WAC can obtain RF load statistics. It should be noted that RF load information will not be immediately used for radio calibration because this process is usually performed every 24 hours. In typical office networks, traffic normally aggregates from 08:00 to 09:00, sinks to a trough from 12:00 to 14:00, and then rapidly increases before plummeting after 18:00.

For this reason, traffic load below a certain threshold is not taken into account. The traffic will not be included in the total amount, and the period of time during which traffic is generated is not counted in the statistical time. For example, in a 24-hour period, an AP has only 6 hours of traffic above the threshold. Therefore, only the traffic of this period is considered. Such a method of obtaining the average traffic in a period of time is referred to as normalized approach.

After radio scanning, 802.11 frame parsing, and spectrum scanning are completed, network topology, channel interference, and traffic load information can be obtained to guide channel and bandwidth assignment. This enables the channel with the least interference and widest bandwidth to be preferentially assigned to radios experiencing the heaviest traffic load.

3.3.2.2 Automatic Transmit Power Control

Transmit power control (TPC) adjusts the transmit power of each radio in networks with multiple APs to prevent the occurrence of coverage holes and overlapping coverage areas. It also dispatches neighbor APs to fill up coverage holes caused by faulty radios. In a coverage hole, STAs cannot receive Wi-Fi signals. If coverage areas mutually overlap, signal interference will occur and roaming experience will be negatively affected.

- If two co-channel APs have a significant overlapping coverage area, co-channel interference will be more severe, decreasing throughput.

- STAs will not roam if they enter the core coverage area of a different AP but still receive strong signals from the currently associated AP, as illustrated in Figure 3.6. Due to asymmetrical uplink and downlink transmit power, signals of the STAs can be rarely received by the associated AP. As a result, user experience deteriorates significantly.

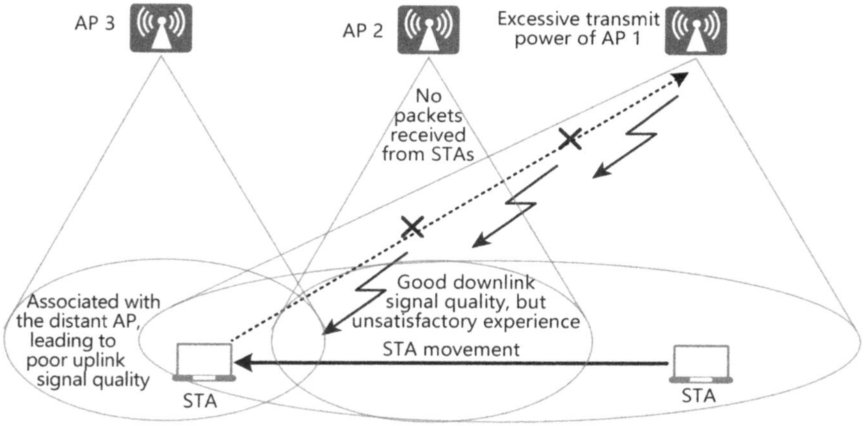

AP 3   AP 2   Excessive transmit power of AP 1

No packets received from STAs

Associated with the distant AP, leading to poor uplink signal quality

Good downlink signal quality, but unsatisfactory experience

STA movement

STA   STA

FIGURE 3.6   STAs still associated with the original AP after moving away, leading to poor experience.

Therefore, TPC is introduced to prevent the existence of an excessively large coverage area while ensuring that there are no coverage holes.

Figure 3.7 illustrates the effect of a TPC algorithm. After radio topology is identified, coverage edges can be determined based on SNR values, and transmit power can subsequently be adjusted. In most cases, it would be the most ideal that the transmit power is the same between a radio and STAs to avoid uplink and downlink asymmetry where STAs can receive packets from the AP but their packets are out of the AP's reach.

In Figure 3.7, the coverage shows an excessively large overlapped coverage area before TPC is performed (left), which reduces the likelihood of STAs roaming and APs listening to the uplink packets of the coverage-edge

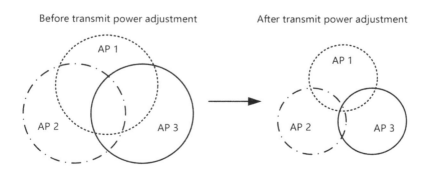

Before transmit power adjustment   After transmit power adjustment

AP 1   AP 1

AP 2   AP 3   AP 2   AP 3

FIGURE 3.7   TPC algorithm effect.

STAs. In this case, user experience is poor. After TPC is performed (right), coverage area overlapping does not exist, and STAs receive high-SNR signals within the AP's coverage area, noticeably improving user experience.

Typically, TPC can also be performed to fill up coverage holes when there are radio faults. As illustrated in Figure 3.8, AP 4 was out of service due to a power failure or abnormal restart because the WAC did not receive its heartbeat packets. With an automatic local radio calibration, AP 2 and AP 3 increase their transmit power, extending coverage to the areas previously covered by AP 4, eliminating the coverage holes. The STAs previously covered by AP 4 now access AP 2 and AP 3, minimizing impact on services.

3.3.2.3 Automatic Band Adjustment

In medium- and high-density networking, APs supporting both 2.4 and 5 GHz bands are generally deployed within close proximity of each other to increase capacity. Due to close proximity and fewer available channels on the 2.4 GHz band, a huge number of APs work on the same channels. This hinders capacity improvement and increases co-channel interference between APs, and it is generally referred to as radio redundancy.

Dual-band APs supporting two 5 GHz channels have been introduced to resolve this issue. These APs are capable of radio switching from 2.4 to 5 GHz so that their radios can work on the 5 GHz channels simultaneously, preventing a large number of APs working on the 2.4 GHz band. This helps increase overall capacity. For APs that do not support radio switching, in the case of noticeable radio redundancy, the 2.4 GHz band can be disabled or work in the monitoring mode, so as to reduce co-channel interference. This is referred to as dynamic frequency band selection (DFBS) or dynamic frequency assignment (DFA).

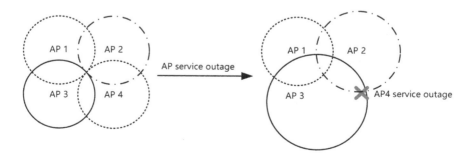

FIGURE 3.8   Filling up coverage holes when there are radio faults.

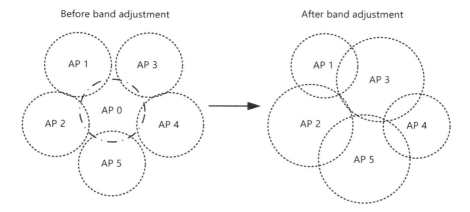

FIGURE 3.9   Adjusting bands.

Figure 3.9 illustrates Huawei's DFBS algorithm. AP 0 is considered a redundant radio if its neighbor APs can increase their coverage to replace it, and then it will be disabled or switched to work on the 5 GHz band or in monitoring mode.

In ultra-high-density WLAN networking, triple-radio APs have been introduced, supporting one 2.4 GHz channel and two 5 GHz channels. In an ultra-high-density network, radio redundancy is likely to occur on both the 2.4 and 5 GHz bands, further increasing interference and deteriorating user experience.

In this case, similar to handling 2.4 GHz radio redundancy, the DFBS algorithm switches redundant 5 GHz radios to work in monitoring mode to detect interference. The radio can also be used for radar detection, eliminating the need for using the AP's other radios for interference and radar detection.

3.3.2.4 Automatic Adjustment of Channels and Bandwidth

Based on obtained neighbor topology, external interference, and long-term load statistics, dynamic channel allocation (DCA) can be further performed to assign channels and bandwidth to APs. The 2.4 GHz band has fewer channels and the bandwidth is normally fixed at 20 MHz. The 5 GHz band is increasingly opened up, providing abundant channels. New 802.11 standards provide better support for high bandwidth. This allows APs to fully utilize channels and increase bandwidth as much as possible, fulfilling throughput requirements. DCA supports dynamic allocation of 5 GHz channel bandwidth in line with network topology, interference, and load statistics.

Automatic adjustment of channels and bandwidth prevents strong interference between adjacent areas, improves user experience, and realizes the potential of high bandwidth. The adjustment is mainly based on neighbor topology, interference, and APs' bandwidth requirements. In networks with hundreds or even tens of thousands of APs, collecting the information and allocating appropriate channels and bandwidth will exponentially increase computing workload.

Huawei's solution to this dilemma is a "division-adjustment" approach. First, the network is divided into small areas, and then channels and bandwidth are adjusted on a per-area basis. The aim of this is to reduce the computing workload by limiting the number of APs for each adjustment. Each adjustment is used as the input for subsequent adjustments. This way, channel bandwidth can be adjusted on the whole network.

Figure 3.10 illustrates how 2.4 GHz channel adjustment is performed in an area. Before adjustment, two APs working on channel 6 are located within close proximity (left), creating strong co-channel interference. After the adjustment, the distance between the two APs noticeably increases (right), minimizing interference on the entire network.

Channel adjustment is different from bandwidth adjustment. With channel adjusted through the DCA algorithm, in a network that has seven APs for providing continuous coverage, the 2.4 and 5 GHz channels are automatically assigned, as illustrated in Figure 3.11. The 2.4 GHz co-channel and adjacent-channel interference has weakened, and the 5 GHz co-channel and adjacent-channel interference has been avoided, showing that the DCA

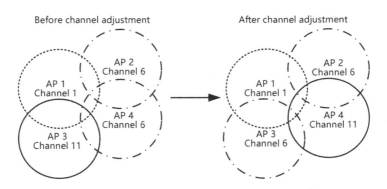

FIGURE 3.10 Channel adjustment.

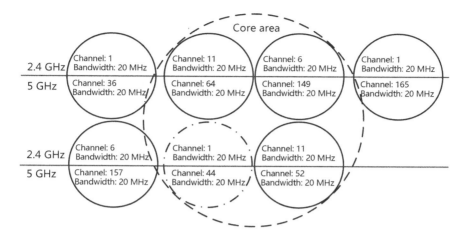

FIGURE 3.11   Effect of the DCA algorithm.

algorithm meets requirements. Historical load information is also used to help allocate channels with low interference to higher-load radios.

Assume that channel 44 in Figure 3.11 has reached its upper capacity limit over the period of time and will continue to fulfill high traffic requirements in the future, and the core coverage areas have significantly higher traffic volume than the coverage edge areas. In this case, the DCA algorithm is no longer applicable. The solution is to add bandwidth optimization to DCA to enable dynamic bandwidth selection (DBS), which is a function that supports bandwidth assignment in line with historical traffic statistics.

Figure 3.12 provides the results of channel and bandwidth allocation on the 2.4 and 5 GHz bands through the DBS function for the seven APs.

As illustrated in Figure 3.12, the bandwidth of channel 44 has changed to 80 MHz. This will allow the service capacity to increase significantly, restricting traffic throughput from reaching the maximum capacity limit. For customers, the core coverage area has improved service capacity without increasing AP density. In addition, the bandwidth of other channels in the core coverage area has increased from 20 to 40 MHz, enabling all 13 channels (per the country code of China) in the 5 GHz band to be fully utilized.

Therefore, with Huawei's automatic channel and bandwidth adjustment technology, interference can be effectively reduced by allocating same-band or adjacent-band resources to APs that are distant from each other. This maximizes the utilization of spectral resources.

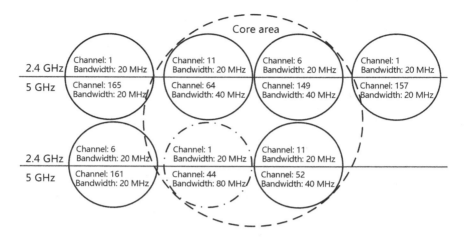

FIGURE 3.12   Effect of the DBS algorithm.

3.3.3 Applications and Benefits

This section describes the applications and benefits of radio calibration. Based on network deployment, maintenance, and modification requirements, Huawei's radio calibration is divided into three types: manual, scheduled, and event-triggered.

3.3.3.1 Manual Radio Calibration

After all APs go online during deployment, radio calibration can be manually triggered so that the network automatically adjusts AP's channel and transmit power, without the need for manual intervention. In networks requiring a huge number of APs, manual channel and transmit power planning is labor-intensive and time-consuming. Radio calibration can be triggered to determine the optimal channels and transmit power for APs, reducing deployment costs while ensuring network quality.

3.3.3.2 Scheduled Radio Calibration

In office areas where many APs are deployed, regular radio calibration is essential for optimal WLAN performance within the context of changing network environments. However, frequent radio calibration will adversely impact normal services. For example, STAs that have a poor compatibility and protocol compliance will possibly be disconnected during channel switching on APs. Therefore, in network O&M, radio calibration is configured to have a large interval or to work at scheduled times, for example, when the traffic reaches the trough point at night. The network collects the traffic statistics, STA quantities, and other information between two moments radio calibration is triggered. This means that, even though the scheduled radio calibration is only triggered at the specified time, high bandwidth can still be allocated to the radios that have high peak-hour traffic.

3.3.3.3 Event-Triggered Radio Calibration

In an office WLAN, over a certain period of time since the network was deployed, the number of STAs accessing the network will keep increasing to the level that existing APs cannot meet coverage and capacity requirements, requiring new APs to be deployed. Automatic radio calibration allows the network to automatically allocate the channel and transmit power to the new APs, eliminating the need for extra planning.

After detecting that a new AP goes online, the WAC delays automatic channel and transmit power allocation until some time later. This is because the AP takes time to detect and collect neighbor information. When the delayed time has elapsed, the DCA and TPC algorithms will be used to calculate the optimal channel and transmit power based on the collected neighbor information. After receiving the calculation results, the AP will start to operate with the allocated channel and transmit power.

Radio calibration is also triggered in the following scenarios:

- An AP is faulty and goes out of service, creating coverage holes in the network. Automatic radio calibration adjusts the transmit power of adjacent APs to fill up the coverage holes. This ensures network reliability and mitigates the out-of-service impact.

- An unauthorized AP interferes with an authorized AP working in the same channel. Automatic radio calibration adjusts the transmit power to reduce or eliminate interference.

- Microwave ovens interfere with an AP. Automatic radio calibration is triggered to reduce or eliminate interference.

Event-based radio calibration is locally performed without affecting the normal operation of the entire network.

Radio calibration is widely used in WLANs, especially in large areas such as universities and hospitals. It plays an important role in ensuring network performance, O&M, and reliability. The benefits of radio calibration include:

- Ensured optimal network performance
  Radio calibration enables intelligent radio resource management in real time, ensuring that WLANs quickly adapt to environmental changes and retain optimal network performance.

- Reduced deployment and O&M costs
  Radio calibration allows for automatic radio management, reducing the skill requirements for O&M engineers and minimizing manual workload.

- Enhanced reliability
  Radio calibration enables automatic monitoring, analysis, and adjustment. This facilitates the response to network performance deterioration and improves reliability as well as user experience.

3.4 STA ROAMING AND SCHEDULING

Although WLAN uses wireless media to improve user mobility, it brings new challenges, such as ensuring that user experience is guaranteed.

As high signal quality is required to guarantee user experience, the communication between STAs and APs using wireless media has become more important. The problem, however, is that the signals between STAs and APs become weaker as their transmission distance grows. To this end, vendors adopt smart roaming technology to enable STAs that move away from an AP to switch to a closer AP, ensuring sustained signal quality. In addition, scheduling helps a WLAN avoid congestion. This section describes how band steering and load balancing help STAs select optimal networks to avoid congestion.

3.4.1 Roaming Overview

3.4.1.1 Technological Background

Signal quality deteriorates as STAs move away from an AP. As such, when the signal quality decreases to a specified level (also known as the roaming threshold), STAs roam to a closer AP to maintain signal quality.

As illustrated in Figure 3.13, the roaming process involves the following actions:

1. While maintaining connected to AP 1, an STA sends probe request frames on all supported channels. Then, after receiving the request on channel 6 (used by AP 2), AP 2 sends a response on channel 6. After receiving the response, the STA evaluates the signal quality of AP 1 and AP 2 and determines the optimal AP to be associated with. In the example provided, AP 2 is selected.

2. As illustrated in 1, the STA sends an association request to AP 2 on channel 6, and AP 2 sends the association response. Then, an association is established between the STA and AP 2.

3. As illustrated in 2, the STA sends a disassociation request to AP 1 on channel 1 (used by AP 1) to dissociate from AP 1.

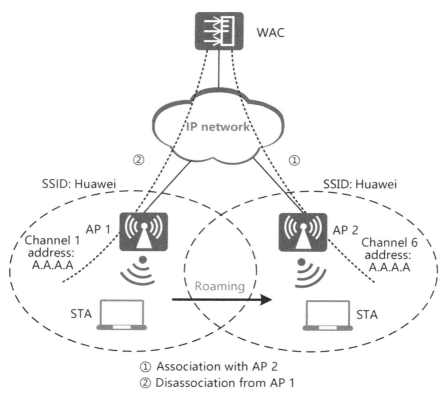

① Association with AP 2
② Disassociation from AP 1

FIGURE 3.13   STA roaming.

If the STA security policy uses open authentication or wired equivalent privacy (WEP) shared key authentication, no further action is required. However, if Wi-Fi Protected Access (WPA) or Wi-Fi Protected Access 2 (WPA2)+pre-shared key (PSK) are used, another step of key negotiation is required. If WPA2+802.1X are used, a further step of 802.1X authentication and key negotiation is required.

Each time an STA accesses an AP, with WPA/WPA2+PSK and WPA2+802.1X, key negotiation or authentication interaction is required. This leads to an extended time spent switching to a new AP and increases risks of service outages and voice and video frame freezing. The following describes how fast roaming and 802.11k assisted roaming are used to solve these challenges.

In addition to authentication, STA stickiness must be considered. Due to inconsistent roaming thresholds among STA vendors, some terminals respond slowly to roaming. As illustrated in Figure 3.14, when moving closer to an AP that provides a higher signal quality, STAs are still associated with the original AP, even when signal quality becomes extremely low. This is referred to as stickiness, while these terminals are referred to as sticky STAs.

Low signal quality causes service experience unable to be guaranteed for sticky STAs. It also decreases data transmission speed for these STAs, leading to more air interface resources consumed than normal. As a result,

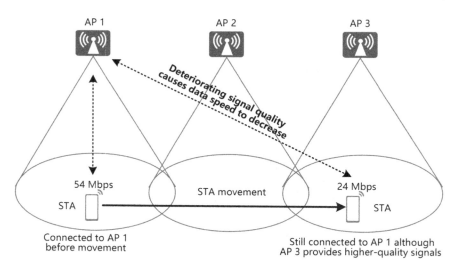

FIGURE 3.14   STA stickiness when roaming to a new AP.

other STAs are adversely impacted. This section expands upon how smart roaming is used to solve this problem.

3.4.1.2 Fast Roaming Principles

Fast roaming simplifies the STA authentication exchange and/or key negotiation process of WPA2+PSK and WPA2+802.1X, shortening the time to roam to a new AP. This section uses Extensible Authentication Protocol-Protected Extensible Authentication Protocol (EAP-PEAP) and Advanced Encryption Standard (AES) as an example to describe the principles of fast roaming.

Based on common roaming to a new AP, the entire packet exchange process includes link authentication, reassociation, STA authentication, and key exchange, as illustrated in Figure 3.15. The last two actions use the majority of the entire exchange time.

Currently, the pairwise master key (PMK) technology and 802.11r fast roaming are used to meet fast roaming requirements.

1. PMK fast roaming

    To improve roaming experience, 802.11i introduces PMK caching to avoid 802.1X authentication. A roaming process with PMK caching is referred to as PMK fast roaming and is only applicable to WPA2+802.1X authentication.

    In PMK fast roaming, STAs include a PMK Identifier (PMKID) in their association/re-association requests. STAs uniquely and irreversibly identify their PMKs through PMKIDs. Then, upon receiving an association/reassociation request, the AP or WAC derives a PMKID based on its own PMK. If the result is consistent with that in the request, the STA is considered having completed 802.1X authentication, and the AP or WAC directly sends the association/reassociation response. The key exchange process continues without another STA authentication process.

2. 802.11r fast roaming

    802.11r, also known as Fast Basic Service Set Transition (FT), was adopted on July 15, 2008. Devices complying with this specification support fast and secure roaming. The roaming mode is referred to as 802.11r fast roaming and is applicable to both WPA2+PSK and WPA2+802.1X authentication.

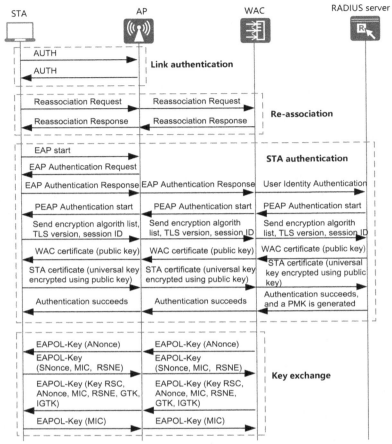

Note: RADIUS stands for Remote Authentication Dial In User Service.

FIGURE 3.15   STA roaming to a new AP.

In 802.11r fast roaming, STA authentication and key exchange are completed during link authentication and association. By doing this, authentication and key exchange are not needed during roaming.

Both roaming technologies require support from both APs and STAs.

1. PMK fast roaming

As described above, PMKID is key to PMK fast roaming.

Upon common roaming or upon the first access to networks with WPA2+802.1X authentication enabled, PMKs are generated on both STAs and the WAC during STA authentication and are used to generate pairwise transient keys (PTKs) during key exchange. A PMK, with STA's MAC address, basic service set identifier (BSSID), and PMK lifetime, forms a pairwise master key security association (PMKSA). A PMKID is generated based on the PMKSA by the formula:

$$PMKID = HMAC\text{-}SHA1\text{-}128(PMK, "PMKName"|BSSID|MAC\_STA)$$

HMAC-SHA1-128 is a hash function. The *PMKID* calculated using this function is unique and irreversible. *PMK Name* is a fixed character string.

Based on this formula, *PMKID* remains unchanged as long as the PMK and MAC addresses are not changed. STAs include the PMKID in their association/reassociation requests. The AP, after receiving an association/re-association request, reports the PMKID to the WAC. Then, the WAC calculates a PMKID and compares the result with the received one. If the two are the same, the WAC considers that the STA has been authenticated and directly instructs the AP to send the STA the association success response. With PMKID successfully verified, key exchange will be directly proceeded without a second authentication.

2. 802.11r fast roaming

802.11r includes the FT protocol and FT resource request protocol, with the latter including an extra resource confirmation packet exchange process on top of the former. With FT resource request procedures, if the destination AP has no resources available for an STA, its roaming access will be denied. Each protocol supports two modes: over-the-air and over-the-DS. In over-the-air mode, packet exchanges are completed only between STAs and the destination AP. By contrast, in over-the-DS mode, the source AP also participates in packet exchanges.

A three-level key system is responsible for key negotiation during association, where the authenticator (AP and WAC) and authentication application ends have their own mechanisms, as illustrated in Figures 3.16 and 3.17.

The following definitions apply in the three-level key system.

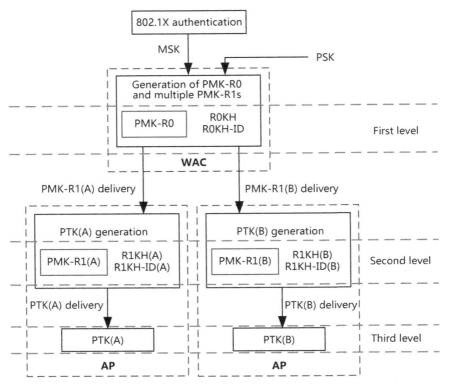

Note: MSK stands for Master Session Key.

FIGURE 3.16 Key mechanism on the authenticator end.

1. PMK-R0

As the first-level key generated based on PSK or 802.1X PMK, PMK-R0 is identified by PMK-R0 Name and saved in the PMK-R0 Key Holder, which is referred to as R0KH and S0KH on the authenticator and authentication application ends, respectively. The R0KH-ID and S0KH-ID are used to identify the R0KH and S0KH, respectively.

2. PMK-R1

PMK-R1 is the second-level key generated based on PMK-R0. It is identified by PMK-R1 Name and saved in the PMK-R1 Key Holder, which is referred to as R1KH and S1KH on the authenticator and authentication application ends, respectively. The R1KH-ID and

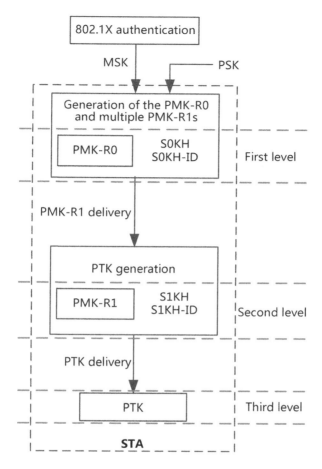

FIGURE 3.17   Key mechanism on the authentication application end.

S1KH-ID are used to identify the R1KH and S1KH, respectively. Multiple PMK-R1s may exist, and the PMK-R1 generator distributes the PMK-R1 to each key holder. In practice, the WAC generates one PMK-R0 as well as multiple PMK-R1s and sends the PMK-R1s to different APs. Each authentication application end includes one PMK-R0, one PMK-R1, and one PTK.

3. PTK

PTK is the third-level key generated based on the PMK-R1. After PTKs are generated at both ends, encrypted communication can be implemented.

As illustrated in the example below, an STA roams from AP 1 to AP 2, while WPA2+802.1X authentication is used on the network side. The roaming diagram is illustrated in Figure 3.18.

1. STA initial access to an FT network

   The WAC notifies STAs of its support for 802.11r by sending-Beacon and probe response frames and broadcasts its own mobility domain identifier (MDID). If an STA supports 802.11r, it replies to the Beacon and probe response frames and initiates the access process.

   First, a common link authentication process is completed.

   Then, the STA sends AP 1 an association request frame containing the MDID. Upon the receipt of the association request frame, AP 1 checks the MDID against its own and replies to the STA with an association response frame. Depending on the check result, if they are consistent, AP 1 sends the association request frame to the WAC,

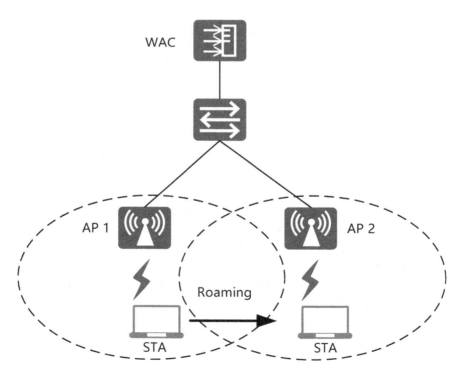

FIGURE 3.18 Fast roaming.

and the WAC replies with an association response frame containing the MDID, R0KH-ID, and R1KH-ID through AP 1. If they are not consistent, AP 1 sends the association response frame indicating that the association is rejected.

The R0KH-ID and R1KH-ID are the MAC address of the WAC and the BSSID of AP 1, respectively, and they are used for the STA to generate PMK-R0 and PMK-R1.

Next, EAP-PEAP authentication is completed between the STA and RADIUS server, after which PMK is generated. The WAC generates PMK-R0 and PMK-R1 that corresponds to AP 1, and sends PMK-R1 to AP 1. PMK-R1 is calculated as follows:

$$PMK\text{-}R1 = KDF\text{-}256 \ (PMK\text{-}R0, \ "FT\text{-}R1", \ R1KH\text{-}ID \| S1KH\text{-}ID)$$

KDF-256 is the key calculation function, *R1KH-ID* is the BSSID, and *S1KH-ID* is the MAC address of the STA.

Lastly, four handshakes are completed between the STA and the WAC, after which the PTK is generated for data encryption. This process is illustrated in Figure 3.19.

2. STA roaming

The following describes how an STA that has established connections to AP 1 in an FT network roams to AP 2.

First, the STA sends AP 2 an authentication request frame containing its MDID, PMK-R0 Name, SNonce, and R0KH-ID.

Next, after receiving the authentication request frame, AP 2 checks whether the MDID is the same as its own MDID. If it is the same, AP 2 extracts the information from the frame and combines the extraction with its own ANonce, R1KH-ID, and STA MAC address before sending it to the WAC.

Then, after receiving the information, the WAC checks whether the PMK-R0 Name is the same as its own. If it is, the WAC generates PMK-R1 and PTK based on the remaining information and sends PMK-R1 and the PTK to AP 2. After receiving the PMK-R1 and PTK from the WAC, AP 2 considers that the STA has passed authentication and sends the STA an authentication response frame containing R1KH-ID, ANonce, MDID, PMK-R0 Name, SNonce, and R0KH-ID.

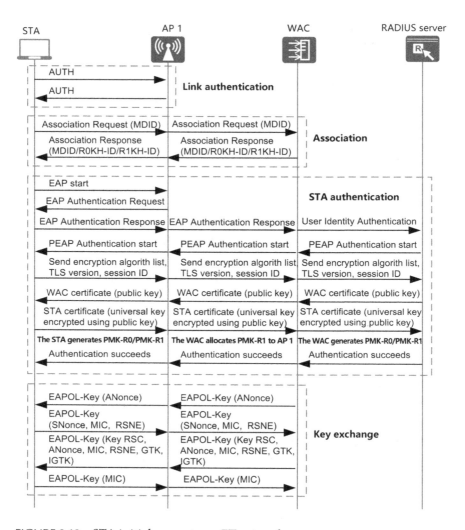

FIGURE 3.19   STA initial access to an FT network.

Lastly, after receiving the authentication response frame, the STA generates PMK-R1 and PTK based on the information contained. Then, the STA completes the authentication for fast roaming.

The STA reassociates with AP 2 through the following process:

First, the STA sends a reassociation request frame containing a message integrity check (MIC). Then, after receiving the request frame, AP 2 uses the generated PTK to verify the MIC, and, if correct, returns a reassociation response frame containing the MIC to

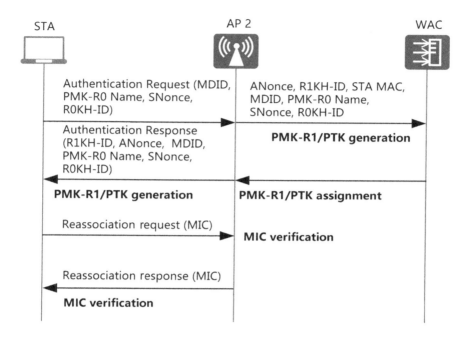

STA   AP 2   WAC

Authentication Request (MDID, PMK-R0 Name, SNonce, R0KH-ID)

ANonce, R1KH-ID, STA MAC, MDID, PMK-R0 Name, SNonce, R0KH-ID

Authentication Response (R1KH-ID, ANonce, MDID, PMK-R0 Name, SNonce, R0KH-ID)

PMK-R1/PTK generation

PMK-R1/PTK generation   **PMK-R1/PTK assignment**

Reassociation request (MIC)

MIC verification

Reassociation response (MIC)

MIC verification

FIGURE 3.20   STA roaming.

the STA. After receiving the response frame, the STA uses the generated PTK to verify the MIC, and, if correct, considers that the reassociation exchange is completed and then disconnects from AP 1.

Then, fast roaming is completed, with the process illustrated in Figure 3.20.

3.4.1.3 Principles of 802.11k Assisted Roaming

802.11k defines the mechanism for measuring WLAN radio resources, enabling STAs and APs to sense radio environment. The following uses Huawei's WLAN products to describe how 802.11k assists in STA roaming (802.11k will not be described in detail).

According to 802.11k, STAs send requests to APs for the list of neighbor APs within the extended service set (ESS). The list contains the BSSIDs and channel numbers, as illustrated in Figure 3.21.

As illustrated in Figure 3.22, after receiving the neighbor AP list containing the channel information of adjacent APs, STAs compliant with 802.11k do not need to search all 2.4 and 5 GHz channels to find roaming destinations. As a result, scanning time and channel scanning overhead

Byte	1	1	6	4	1	1	1	variable
	Element ID	Length	BSSID	BSSID information	Operating class	Channel number	PHY type	Optional subelements

FIGURE 3.21   Information about neighbor APs.

Signal strength: −70 dBm
When needing to roam, scan all channels 36, 40, 44, 48, 52, 56, 60, 64, 149, 153, 157, 161, 165...

Total time: 6s

802.11k-incompliant

Signal strength: −70 dBm
When needing to roam, scan optimal channels 40, 48, and 157.
Is suitable AP found? If yes, roam to the AP.

Total time: 200 ms
No suitable APs found? Scan all channels.

802.11k-compliant

FIGURE 3.22   Principles of 802.11k assisted roaming.

are both reduced. A reduced scanning time ensures improved roaming experience, while reduced channel scanning times minimize unnecessary channel switching and probe requests, prolonging the battery life of STAs.

As illustrated in Figure 3.23, STA-obtained neighbor AP lists are dynamically generated. Therefore, neighbor AP lists are different among STAs connecting to different APs.

3.4.2 Smart Roaming

3.4.2.1 Principles of Smart Roaming

Although sticky STAs only account for a minority of STAs, they pose enormous risks to networks.

1. Decreased throughput

    STAs roam to APs for a higher-quality signal coverage that enables a higher data rate. However, STA stickiness restricts STA roaming, forcing STAs to exchange data at a low rate. As such, longer times are needed for them to occupy the air interface resources, preventing other STAs (particularly high-speed ones) from achieving a higher throughput, while also decreasing the throughput of the AP.

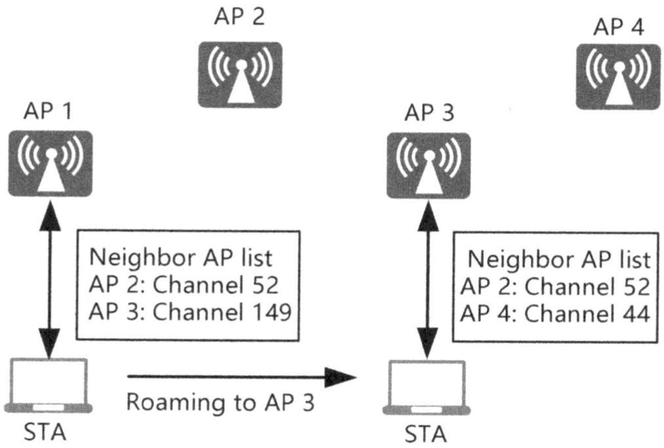

FIGURE 3.23   Obtaining different neighbor AP lists when accessing different APs.

2. Impacted user experience

Sticky STAs do not switch to other APs that provide higher signal quality. As a result, the coverage deteriorates, decreasing data rate and impacting user experience.

3. Disrupted channel planning

In networks with multiple APs, channel planning is performed to maximize network capacity by reducing interference through channel multiplexing. However, STA stickiness restricts channel planning because it leads to a channel outside of an area being used, possibly inviting interference with the outside channel. As a result, capacity decreases, affecting user experience.

Smart roaming is used to identify and guide sticky STAs to more suitable APs, eliminating adverse impacts. The smart roaming process is as follows:

1. Step 1: Identifying sticky STAs

Based on network planning, each AP has a core coverage area, with coverage edges determined according to edge signal strength required by edge throughput. In the example illustrated in Figure 3.24, the edge signal strength is −65 dBm, and each service area is covered by at least one AP (coverage areas overlap when multiple APs provide coverage). However, if the signals received by an

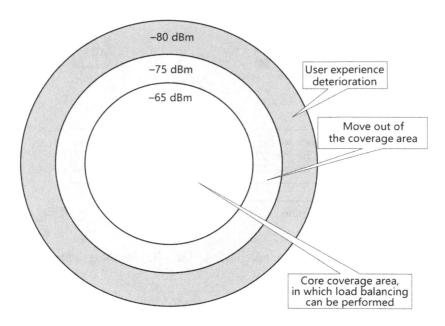

FIGURE 3.24   Coverage areas and signal strength.

AP from an STA are weak when compared to the AP's edge signal strength, the STA has entered the coverage area of another AP, and STA roaming will be triggered as a result. As such, STA stickiness can be determined based on the signal strength received on APs with which STAs are associated.

2. Step 2: Collecting information about neighbor APs

To help an identified sticky STA select a more suitable AP, the network collects information about adjacent APs through measurement. 802.11k provides a mechanism for measuring adjacent APs and collecting related information. For STAs supporting 802.11k, measurement is completed using this mechanism. For STAs not supporting 802.11k, APs perform active scanning and listening to implement measurement.

For an identified sticky STA supporting 802.11k, upon detecting a stickiness behavior, the associated AP instructs the STA to perform neighbor measurement based on 802.11k, while the STA sends the AP the measured signal strength of adjacent APs based on the 802.11k notification mechanism, as illustrated in Figure 3.25.

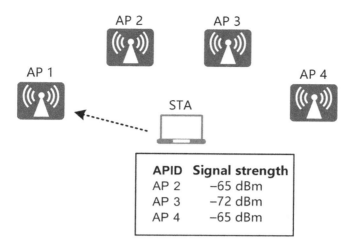

FIGURE 3.25   Notification mechanism of 802.11k.

For an identified sticky STA not supporting 802.11k, the network uses the data obtained by AP scanning. The data is generally a summary of the scanning results of all APs within a period of time. However, as AP scanning is completed at different times, the information will be, more often than not, misleading for quickly moving STAs, as illustrated in Figure 3.26.

To address the challenge of selecting the most appropriate destination AP for STA roaming, a collaboration mechanism between APs has been introduced to help APs notify each other of their load and signal strength, as illustrated in Figure 3.27. However, the coordination is based on AP channel switching and scanning. On the APs processing voice, video, or other delay-sensitive services, the scanning process is not performed to ensure service experience.

3. Step 3: Selecting an appropriate AP for sticky STAs

Based on 802.11k scanning (for supporting STAs) or the information collected through AP collaboration (for not-supporting STAs), the network generates a neighbor AP list for sticky STAs and selects a destination AP from the list for each of the STAs to roam to. The destination AP must meet the following requirements:

• Load balancing can be achieved. For details, see Section 3.4.4.

• Preferentially, the 5 GHz band is used. For details, see Section 3.4.3.

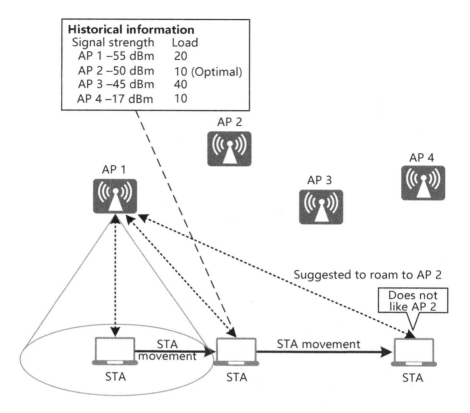

FIGURE 3.26 Information becoming misleading for quickly moving STAs.

The network selects the AP providing the strongest signal while meeting also the requirements as the destination AP.

4. Step 4: Roaming sticky STAs to the destination AP

802.11v extends the network equipment management function, with a mechanism added for APs to instruct STAs to roam. The process is as follows:

First, the AP sends the STA a BSS Transition Management (BTM) request frame. The information about the recommended destination AP for roaming can be contained in the frame.

Then, after receiving the BTM request frame, the STA determines whether to roam to the destination AP and notifies the AP of its decision in the BTM response frame. Figure 3.28 illustrates the process of the STA determining to roam to the recommended destination AP.

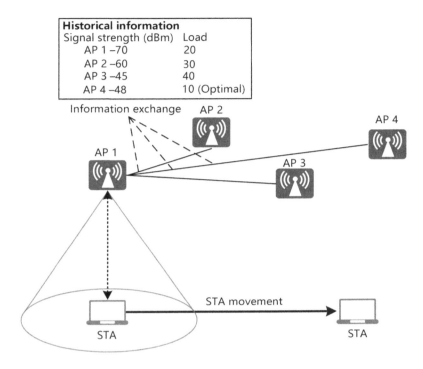

FIGURE 3.27   Selecting an appropriate destination AP through AP collaboration.

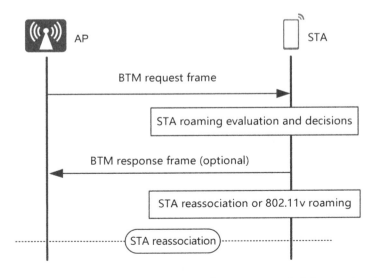

FIGURE 3.28   STA determining to roam to the recommended destination AP.

3.4.2.2 Applications of Smart Roaming

Smart roaming is introduced to address STA stickiness. Although sticky STAs account for a minority of STAs, they pose enormous risks to networks. In particular, in dense networking environments, sticky STAs impact WLAN performance even if there are few isolated cases.

Therefore, smart roaming is used in any scenarios where sticky STAs are present, particularly in open offices where employees' work stations are not fixed and in large public places having heavy flows of mobile traffic, such as airports and railway stations.

For example, in mega conferences for product releases where free high-quality Wi-Fi services will be provided, sticky STAs can regularly compromise service experiences of both guests and media reporters. After leaving from one area to another, some guests found that the Wi-Fi network at the new meeting site was slow or unavailable even though APs are seen here. In fact, with a second check on their devices, these guests found signal bars indicated too weak signals. In this example, this issue is attributed to STA stickiness, making their devices sticky to the original APs rather than roam to new APs. With smart roaming, such sticky STAs are promptly recognized and the optimal approaches will then be used to help them roam to the new APs to continue the services. As a result, the adverse impact caused by sticky STAs can be mitigated or eliminated, improving user experience and capacity usage.

Huawei has also introduced smart roaming to industrial applications. For example, automated guided vehicle (AGV) roaming in warehousing is used to enable data transmission featuring an almost zero packet loss. This makes it possible to ensure non-stop AGV operation, substantially improving the efficiency.

3.4.3 Band Steering

3.4.3.1 Technological Background

Currently, the spectrum resources designated for WLAN technology include:

- 2.4 GHz band, with a total bandwidth of 83.5 MHz between 2.4 and 2.4835 GHz, as illustrated in Figure 3.29 On this band, there are few differences between countries.

- 5 GHz band, which is an unlicensed national information infrastructure (UNII) band. On this band, the usable resources include

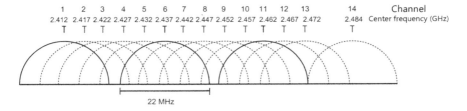

FIGURE 3.29   Spectrum resources on the 2.4 GHz band.

UNII-1 80 MHz, UNII-2 80 MHz, UNII-2 Extended 220 MHz, UNII-3 80 MHz, and ISM 20 MHz, as illustrated in Figure 3.30. This band provides much more spectrum than 2.4 GHz, while there are also considerable differences among countries. For example, China does not open up the UNII-2 Extended part, whereas South Korea only opens up the UNII-3 part.

In addition to the substantial difference in the amount of spectrum resources, the 2.4 and 5 GHz bands differ greatly in terms of interference and standard support.

- As the 5 GHz band has abundant spectrum, 802.11ac supports the aggregation of multiple 20 MHz channels into 80 MHz or even 160 MHz channels. Higher-bandwidth channels accommodate a larger amount of data traffic than low-bandwidth channels. For STAs, a higher data transfer rate results in shorter file and video download time.

- The 2.4 GHz band is the first spectrum designated for Wi-Fi networks and has few differences among countries. Many common

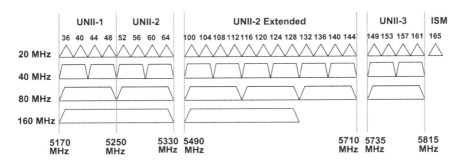

FIGURE 3.30   Spectrum resources on the 5 GHz band.

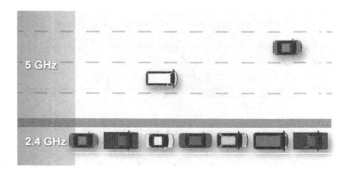

FIGURE 3.31   Difference between the 2.4 and 5 GHz bands.

devices, such as Bluetooth, microwave ovens, and monitoring and remote control devices, all operate on this band. Therefore, this band has limited resources but is mostly used, increasing congestion.

Figure 3.31 illustrates the difference between the 2.4 and 5 GHz bands using an analog. The 2.4 GHz band is like a narrow county road, filled with various obstacles that slow down vehicle movement. By contrast, the 5 GHz band is like a wide, open highway on which high speeds are possible. For dual-band STAs (supporting both the 2.4 and 5 GHz bands), the 5 GHz band is certainly the optimal choice for Wi-Fi services in most cases.

Currently, not all dual-band STAs select the optimal 5 GHz band. Out-of-date devices select networks only based on signal strength, and they do not actively camp on the 5 GHz band. Despite the 5 GHz preference function being implemented on smartphones by some terminal vendors, their devices still access the 2.4 GHz band in some cases.

Huawei has developed band steering as a supplementary solution to STA uncontrollability over band selection in WLANs. This solution helps steer STAs to the 5 GHz band, increasing the proportion of STAs accessing the 5 GHz band.

3.4.3.2 Band Steering Principles

Band steering involves dual-band STA identification and follow-up 5 GHz band steering.

1. Dual-band STA identification

To discover networks, STAs send probe request frames to APs and receive probe response frames from the APs. Dual-band STAs send probe request frames on all channels in 2.4 and 5 GHz bands. As illustrated in Figure 3.32, if an AP receives probe request frames in both 2.4 and 5 GHz bands, the STA is a dual-band terminal.

2. 5 GHz band steering

Band steering is performed both in and after the association phase.

1. In-association steering: As mentioned above, STAs discover networks by sending probe request frames, which is referred to as active scanning. As illustrated in Figure 3.33, APs temporarily do not send STAs probe response frames in the 2.4 GHz band, steering STAs to the 5 GHz band.

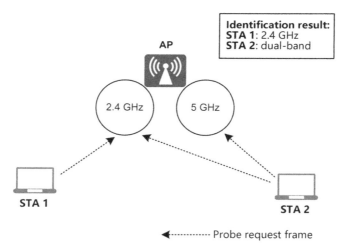

FIGURE 3.32   Identifying dual-band STAs.

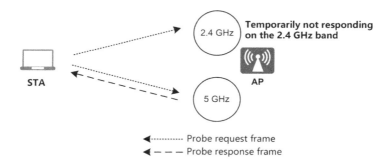

FIGURE 3.33   In-association steering.

2. After-association steering: In addition to active scanning, STAs use a passive method that involves listening to Beacon frames sent by APs to discover networks. For STAs that still access the 2.4 GHz band when no probe response frames are received during active scanning, the APs steer them to the 5 GHz band based on 802.11v.

Figure 3.34 illustrates an AP sending a BTM Request to notify an STA of roaming based on 802.11v.

3.4.3.3 Applications of Band Steering

Band steering depends on the abundant and high-quality spectrum resources of the 5 GHz band.

Due to the easy availability of the 5 GHz channel resources, band steering provides clear benefits to networks. To reach the full potential of the abundant 5 GHz spectrum, the proportion of STAs that are steered to the 5 GHz band is set to 90% by default. In countries where the 5 GHz spectrum is not widely available, such as South Korea (where only four UNII-3 20 MHz channels are usable), the steering proportion can be adjusted. If interference is severe on the 2.4 GHz band, it is a good practice to steer STAs to the 5 GHz band.

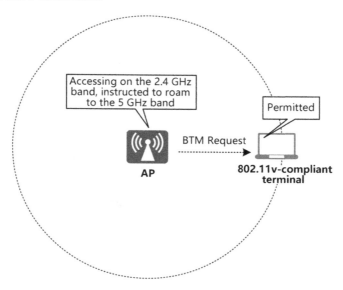

FIGURE 3.34   802.11v band steering.

Band steering can be used typically in the following scenarios:

1. Offices and meeting rooms

   STAs include personal smartphones, personal office devices (such as laptops and tablets), printers, and fax machines. The majority of personal smartphones and office devices support dual bands. Band steering is used to enable access to the 5 GHz band for high-speed services.

2. Exhibition halls

   As large exhibition halls are generally filled with traffic-hungry STAs, a large number of APs are deployed to ensure quality network services. As described above, air interface resources are shared in WLANs, and due to spectrum limitation on the 2.4 GHz band, co-channel interference will be aggregated in dense networking. In addition, Bluetooth devices and other electronic devices will create additional interference, further deteriorating user experience on this band. To solve this issue, steering STAs to the 5 GHz band offers the optimal solution.

3.4.4 Load Balancing

3.4.4.1 Technological Background

As mentioned above, STAs are like vehicles, whereas WLAN is like a highway. By using new technologies, car capabilities are increased significantly in terms of avoiding congestion. However, WLAN STAs are still not as that powerful to avoid congestion by themselves. This section will describe how the network handles traffic congestion for STAs in WLANs.

Based on current Wi-Fi standards, though STAs can be aware of and obtain AP load, they are still associated with heavy-load APs in most cases. As a result, the following impact on services will arise.

1. An excessive number of STAs access the same AP over the same channel, leading to an insufficiency of air interface resources. As a result, STAs must repeatedly contend for chances to send packets based on WLAN communication mechanisms, wasting even more air interface resources and decreasing AP throughput. An increased number of STAs results in more resources wasted. As illustrated in Figure 3.35, with 50 STAs, throughput decreases by up to 30%–50%.

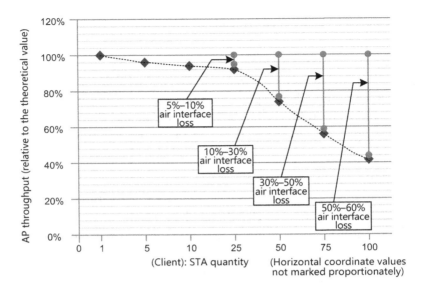

FIGURE 3.35 Mapping between STA quantity and AP throughput.

2. With channel resources shared, as the number of STAs increases, the contention not only wastes air interface resources but also decreases the resources available for each STA.

To solve this problem, load balancing algorithms are introduced in WLANs to improve scheduling.

3.4.4.2 Load Balancing Principles

Load balancing enables the network to instruct STAs to select suitable APs based on load. It includes active load notification and active load balancing.

1. Active load notification

802.11e defines a clear set of load notification mechanisms, allowing networks to include the **QBSS Load** IE in Beacon frames and probe response frames. This IE is used to indicate the number of online STAs connected to APs and their channel utilization information, helping STAs make decisions based on load information. As mentioned above, current load-based decisions are not satisfactory and therefore are only auxiliary to the load-balancing function. In the future, load-based decisions will be more intelligent on STAs, ultimately delivering more benefits.

2. Active load balancing

In WLANs, before selecting APs, the network first checks whether an AP is overly loaded. If it is, the network determines which APs are available to provide resources and how likely these alternatives are to experience heavy load.

The active load balancing process is described below:

1. Evaluate AP load

In WLANs, load is measured based on two indicators:

- STA quantity: This indicator provides static load measurement. That is, whether an AP is heavily loaded depends on how much traffic the STAs associated with it are generating. If its STAs do not generate high levels of traffic, the AP is lightly loaded. In the event of an excess of traffic on each STA, the AP will be considered heavily loaded even if the number of STAs is relatively small.

- Channel utilization: This indicator provides dynamic load measurement, which is more accurate than static load measurement. A WLAN, unlike in wired networks, relies on air interfaces for data transmission. Air interfaces cannot be used unless the request is permitted and the use is limited within a certain period of time. Channel utilization indicates how busy an AP's air interfaces are and therefore reflects its actual load. In addition, as traffic volume changes with services and leads to changes in channel utilization, channel utilization dynamically reflects the load of an AP.

- In WLANs, load is most commonly based on STA quantity, even though it can be evaluated by both STA quantity and channel utilization. On the one hand, after carrying out sufficient observations, it can be concluded that there is a striking similarity among STA services, confirming a positive correlation between the number of STAs and channel utilization. That is, APs having a larger STA quantity will become more heavily loaded. On the other hand, however, because channel utilization is directly affected by service changes, if it is the factor mainly used for evaluating load, STAs will be frequently steered to other APs, increasing the likelihood of service deterioration. The following mainly describes load balancing based on STA quantity.

- Load evaluation based on STA quantity adopts a comparative approach. This means that if an AP has a heavier load than its adjacent APs, STAs served by the AP can be migrated to its adjacent APs. As such, this approach achieves load balancing by evenly distributing STAs to different APs.

2. Select a candidate AP for STAs

For STAs, a candidate AP for load balancing must provide sufficient signal strength to service requirements and must be lightly loaded. Signal strength is a key point of load balancing.

In WLANs, two indicators are used to reflect AP signal strength.

- Downlink signal quality: Indicates the signal strength derived from the AP-to-STA packet reception.

- Uplink signal quality: Indicates the signal strength derived from the STA-to-AP packet reception.

As illustrated in Figure 3.36 uplink signal quality is different from downlink signal quality. This requires both uplink and downlink signal quality to meet network requirements (received signal strength indicator, or RSSI, is not smaller than −65 dBm or SINR

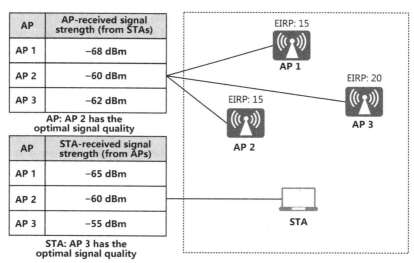

Note: EIRP stands for equivalent isotropically radiated power (EIRP).

FIGURE 3.36 Difference between uplink and downlink signal quality.

not smaller than 30 dB). Otherwise, overall service experience cannot be guaranteed.

To collect complete information in the process of selecting the most suitable AP for load balancing, the network obtains downlink signal quality by using the notification mechanism of 802.11k, while also gathering the information about uplink signal quality, as illustrated in Figure 3.37.

Therefore, the network selects a candidate AP for load balancing by taking into account of both load and uplink and downlink signal quality.

3. Steer STAs to candidate APs

For details, see Section 3.4.2.

3.4.4.3 Applications of Load Balancing

Load balancing is applicable to scenarios where multiple APs are deployed to provide continuous coverage and meet high-performance requirements and coverage areas among APs are overlapped. Load balancing is used in open office areas, large meeting rooms, lecture halls, and exhibition halls. In cases where APs are deployed to mainly ensure Wi-Fi coverage, for example, outdoor APs are deployed to ensure wireless coverage for Wireless City, load balancing is not suitable.

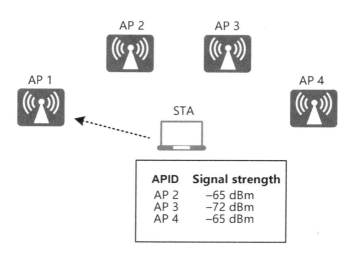

FIGURE 3.37 Notification mechanism of 802.11k.

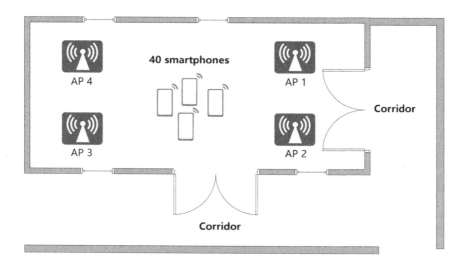

FIGURE 3.38   Load balancing in a meeting room.

1. Meeting room

As illustrated in Figure 3.38, in a large meeting room where multiple APs are deployed, load balancing can be adopted to evenly distribute STAs to these APs. This helps prevent user experience from being affected as a result of an excessive number of STAs accessing only one AP. Table 3.2 illustrates the effect of load balancing.

2. Stadiums with heavy flows of data traffic

It has been a growing trend to deploy wireless networks in large sports stadiums. For example, Huawei has already deployed wireless networks in Europe for Borussia Dortmund and AFC Ajax. These wireless networks play an important role in facilitating networks' access for fans, game broadcasts, and media reports. As illustrated in Figure 3.39, in a user-crowded stadium, the 5 GHz band is used to ensure access requirements of tens of thousands of

TABLE 3.2   Load Balancing Effect

STA Type	Number of STAs Accessing AP 1	Number of STAs Accessing AP 2	Number of STAs Accessing AP 3	Number of STAs Accessing AP 4
Smartphone	10	10	10	10

FIGURE 3.39   Stadiums with heavy flows of data traffic.

users. The APs are deployed with a minimum spacing of 7 m, and they all operate in dual-band mode. This way, load balancing is used to steer dual-band STAs to the 5 GHz band, maximizing the number of STAs accessing each AP and improving the network's the overall capacity.

3.5 ANTI-INTERFERENCE TECHNOLOGY

Wi-Fi frequency spectrum is clustered around unlicensed ISM bands, which are open to all users and devices without incurring spectrum usage costs. This leads to a variety of devices working on the bands, placing WLANs under ubiquitous interference. Although radio calibration helps avoid interfered channels, as channels are limited in WLANs, interference is still an important factor affecting the air interface performance, particularly in networks with multiple APs in a small area.

Interference leads to signal reception failures and prolongs packet exchange delays, while also misleading Wi-Fi devices to considering that their channels are always busy. Consequently, overall data transmission becomes less efficient, Wi-Fi network performance is compromised, and user experience cannot be guaranteed. To mitigate the impact of interference on the air interface performance, anti-interference technology is specially introduced to Wi-Fi networks.

WLAN interference is classified into non-Wi-Fi interference and Wi-Fi interference. Based on interference sources, diversified anti-interference technologies are developed.

Non-Wi-Fi interference is generated by signals from sources not complying with 802.11 standards, such as Bluetooth, electronic shelf labels, ZigBee, cordless phones, baby monitors, microwave ovens, wireless cameras, and infrared sensors. Spectrum analysis is a common approach to locate non-Wi-Fi interference sources in networks, helping users clear the interference sources discovered in networks.

Wi-Fi interference is generated by signals from the sources complying with 802.11 standards. Wi-Fi interference is further categorized as co-channel and adjacent-channel interference, depending on working frequencies. The former occurs between Wi-Fi devices working on the same channel and the latter arises between adjacent channels. Clear channel assessment (CCA) and Request To Send / Clear To Send (RTS/CTS) are mainly used to minimize the probability of collision between desired signals and interfering signals. The adaptive modulation and coding (AMC) algorithm is used to reduce signal loss in cases where interference cannot be averted.

3.5.1 Spectrum Analysis

3.5.1.1 Technological Background

The collision between non-Wi-Fi signals and desired Wi-Fi signals causes packet parsing failures, adversely affecting WLAN user experience. Spectrum analysis allows non-Wi-Fi interference sources to be quickly and effectively detected. After an interference source is detected, alarms will be reported, with information about types, channels, strength, and duty cycles provided to facilitate locating operations. This way, clearing the interference source becomes easier.

3.5.1.2 Huawei's Spectrum Analysis Principles

Figure 3.40 illustrates Huawei's WLAN spectrum analysis solution. It consists of a spectrum sampling engine, spectrum analyzer, and interference

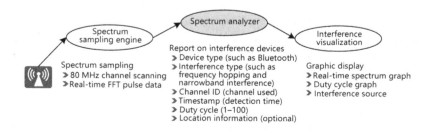

FIGURE 3.40 Spectrum analysis architecture.

visualization. APs incorporate the functions of spectrum sampling engine and spectrum analyzer, and the spectrum drawing server completes the interference visualization function.

1. Spectrum sampling

The spectrum sampling engine periodically scans the air interface to capture radio signals in the time domain, converts the signals from the time domain to the frequency domain through FFT, and samples the converted signals.

In WLANs, spectrum sampling is completed by the spectrum sampling engine of APs. Two scanning modes are supported: hybrid and monitoring mode. In hybrid mode, spectrum-sampling resources are used to transmit service data predominantly unless spectrum sampling is performed on the working channel, and both scanning interval and duration are configurable. Comparatively, in monitoring mode, spectrum-sampling resources are used only for spectrum sampling to ensure real-time data collection. Generally, channels are periodically switched to ensure that the entire band is scanned, with fixed scanning interval but configurable duration.

APs perform real-time FFT operations on the collected radio signals. In hybrid mode, a 20 MHz channel is divided into 64 subcarriers, with a spacing of 312.5 kHz (according to Wi-Fi 5 standards or earlier). Fifty-six of the subcarriers are used for signal transmission, and spectrum sampling results (56 FFT bins) are generated from the signals received on the 56 subcarriers.

In monitoring mode, an 80 MHz channel is divided into two 40 MHz sub-channels. Each 40 MHz channel is divided into 128 subcarriers, with a spacing of 312.5 kHz. Spectrum sampling results (128 FFT bins) are generated from the signals received on the 128 subcarriers.

2. Spectrum analysis

In spectrum analysis, pulse identification and pulse matching are performed to obtain basic pulse information. After statistical analysis and feature extraction of the pulse information, the feature values of interfering signals are obtained, including time signature, center frequency and bandwidth, spectrum symbol difference, duty cycle, pulse symbol difference, pulse off-time symbol difference, and pulse extension. The feature values are then matched against known

non-Wi-Fi interference feature values in the feature database to identify the type of non-Wi-Fi device.

a. Pulse identification

Each sample is subject to pulse identification to calculate the
center frequency and bandwidth of each pulse. The calculation is
performed on a basis of per sampling point. If a sampling point
includes multiple pulses, all the pulses must be identified.

- Calculate the power value of each FFT bin and search for the
local maximum peak point of the power value. Then, choose
the peak points where the power value exceeds the minimum
threshold.

- For each peak point meeting the preceding conditions, from
the FFT bins whose power value exceeds the minimum
threshold but is less than the peak-point power value by no
more than the specified value, choose adjacent continuous
FFT bins to construct a pulse.

- Calculate the center frequency (k_c) and bandwidth (B) based
on the formulas below:

$$k_e = \frac{1}{\sum_k p(k)} \sum_k kp(k), k'_s \leq k \leq k'_e$$

$$B = 2\sqrt{\frac{1}{\sum_k p(k)} \sum_k (k-k_c)^2 p(k)}, k'_s \leq k \leq k'_e$$

b. Pulse matching

The pulse recognition results are stored in an array of sampling
points. During pulse matching, the pulses of each sampling point
are traversed in sequence based on sub-channels. The pulses
having the same feature values form a pulse sequence. The pulse
matching module is used to generate two lists: the active pulse list
and the completed pulse list.

- During scanning on the first sub-channel, the two lists are
empty.

- Upon receiving a pulse, if the active pulse list is empty, the pulse detection module directly adds the pulse to the active pulse list.

- Upon receiving the next pulse, the pulse detection module compares it with the pulses present in the active pulse list. If two pulses match, the module combines the new pulse with the matched one to increase the pulse duration and then calculates the weighted average power value. Two pulses are considered matched when having the same center frequency and bandwidth but power values being no more than 3 dB smaller than the peak-point power value.

- When the time interval of two data samples is excessive (greater than 150 µs), the pulse detection module moves all pulse entries in the active pulse list to the completed pulse list.

- When the sub-channel scanning ends, the pulse detection module transfers all pulses to the completed pulse list.

c. Pulse statistics

After pulses are analyzed, multiple pulse lists are generated based on all sampling points in a sampling period. Each pulse list corresponds to a sub-channel. All data of a sub-channel is processed to calculate the following:

- Average power: Average value of the power of all FFT bins sampled on the sub-channel.

- Average duty cycle: Average value of the duty cycles of all FFT bins sampled on the sub-channel. (For an FFT bin, its duty cycle is 1 when its power value is above the minimum threshold and 0 otherwise.)

- High duty-cycle area: The center frequency and bandwidth are calculated based on the average power in the same way peak value detection is performed to identify areas with high duty cycles.

d. Feature extraction

Feature extraction is performed on the identified pulse sequence to help device identification. The feature values to be extracted mainly include center frequency and bandwidth, spectrum symbol difference, duty cycle, pulse symbol difference, pulse off-time symbol difference, and pulse extension. The pulse extension refers to the average value and variance of pulse quantity of multiple sub-channels, as well as pulse distribution, and it is used only in nonworking mode). Device-peculiar features (such as those related to frequency scan) will also be extracted.

e. Device identification

The extracted feature values are matched against devices in the device list. If a match is found, a non-Wi-Fi device is considered existent. If no match is found, an unknown non-Wi-Fi interference source is recorded.

3. Interference visualization

APs upload the spectrum data to the spectrum drawing server where interference information is visualized to create spectrum graphs.

3.5.1.3 Applications of Spectrum Analysis

Spectrum analysis is predominantly used to detect non-Wi-Fi interference sources in WLANs. In radio environments with various non-Wi-Fi interference devices, spectrum analysis is useful. Applicable scenarios include campus offices (which have cordless phones, asset positioning devices, wireless printers, and other devices), large shopping malls and supermarkets with a large number of electronic shelf labels, and outdoor scenarios with high quantities of uncontrollable non-Wi-Fi devices.

With spectrum analysis enabled, APs upload interference features to the spectrum drawing server where visualized spectrum graphs are created. This allows users to intuitively understand the channels, strength, and types of interference sources. The spectrum analysis results provide an important reference to clearing non-Wi-Fi interference sources and adjusting AP channels. They are also useful in selecting the appropriate anti-interference technologies to reduce the impact of interference signals on WLAN performance.

3.5.2 Clear Channel Assessment

3.5.2.1 Technological Background

Spectrum analysis works only for partial non-Wi-Fi interference. To help analyze Wi-Fi interference and uncontrollable non-Wi-Fi interference, 802.11 introduces a CCA mechanism to monitor channel status (busy or clear). Channel contention occurs between Wi-Fi devices only when chan-nels are clear. With the CCA mechanism, signal transmission in the pres-ence of interference can be avoided to ensure WLAN performance.

3.5.2.2 Working Principles

1. CCA mechanism

WLAN data transmission follows a "listen-to-send" principle. Before sending data, a Wi-Fi device does not prepare to send data until detecting that the channel is clear. The technology used is referred to as CCA, which detects whether a channel is clear at the physical layer. Figure 3.41 illustrates the CCA definition of the IEEE. Basically, CCA uses two thresholds to determine whether a channel is clear: signal detect threshold and energy detect threshold.

Signal detect threshold: On a 20 MHz channel, if the strength of a preamble-contained OFDM signal is greater than or equal to −82 dBm, carrier sensing considers that the channel has a 90% probabil-ity of being busy for 4 µs.

Energy detect threshold: On a 20 MHz channel, if signal strength is greater than or equal to −62 dBm (signal detect threshold plus 20 dBm), carrier sensing considers that the channel is busy.

Based on IEEE definitions, when the channel bandwidth is 20 MHz, the signal detect threshold is equal to the receiver sensitivity of the smallest modulation and coding scheme (MCS) index (which is −82 dBm), and the energy detect threshold is the sum of the signal detect threshold plus 20 dBm, which is therefore −62 dBm. When the channel bandwidth is 40 MHz, the signal detect threshold is −82 dBm for the primary 20 MHz channel and −72 dBm for the second-ary channel. In the case of the channel bandwidth being 80 MHz, the

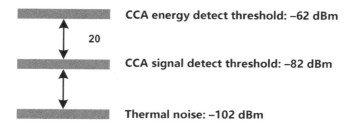

FIGURE 3.41   CCA definition.

signal detect threshold is −79 dBm for the primary 40 MHz channel and −72 dBm for the secondary channel.

To further improve CCA, IEEE 802.11ax introduces the BSS color technology. As illustrated in Figure 3.42, with BSS color recognizing the Intra-BSS (MYBSS) and Inter-BSS (OBSS) frames, it is possible to raise the CCA signal detect threshold for the Inter-BSS frames while maintaining a low CCA signal detect threshold for the Intra-BSS frames. This way, when the signal strength from a neighboring BSS exceeds the CCA signal detect threshold used in earlier standards, with the transmit power lowered for the packets to be sent, the channel can be still considered clear and available for new data transmission. This will minimize the impact of co-channel interference on WLAN performance.

2. Huawei's dynamic CCA mechanism

Packets are sent only when channels are clear. The CCA mechanism monitors channel status (clear or busy) to reduce the likelihood of collision occurring due to unknown channel status. The mechanism is applicable where both non-Wi-Fi interference and Wi-Fi interference are present. It improves the accuracy of channel status detection based on the signal detect threshold, reducing air interface collision and improving transmission efficiency. However, with the same default signal detect threshold, the CCA mechanism cannot deliver the same performance in all scenarios.

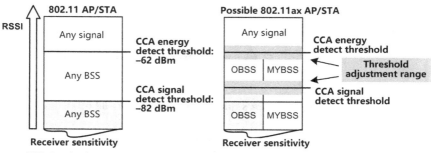

FIGURE 3.42   CCA based on BSS color.

Huawei's dynamic CCA mechanism dynamically adjusts the signal detect threshold based on scenario conditions, maximizing its network performance.

- Lower the CCA signal detect threshold for reduced probability of collision

  As illustrated in Figure 3.43, AP 1 and AP 2 are co-channel APs, use the default CCA signal detect threshold (−82 dBm), and are unaware of each other. When AP 1 sends packets to STA 1, AP 2 considers that the channel is clear based on the CCA mechanism and then sends packets to STA 2. As a result, packets from AP 1 cannot be parsed on STA 1.

  With Huawei's dynamic CCA mechanism, the channel collision as a result of using the default CCA signal detect threshold can be recognized. In case of a high channel collision rate, the CCA signal detect threshold can be adjusted downward to reduce the probability of collision. As illustrated in Figure 3.44, after the CCA signal detect threshold is adjusted to −85 dBm, AP 1 and AP 2 become mutually detectable. Therefore, when AP 1 sends packets to STA 1, AP 2 considers that the channel is busy based on the CCA mechanism and will not send packets to STA 2 until the channel becomes clear. This reduces the probability of channel collision.

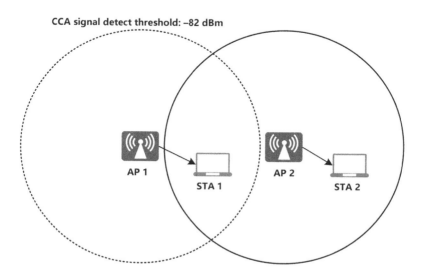

FIGURE 3.43   Scenario 1—default CCA signal detect threshold.

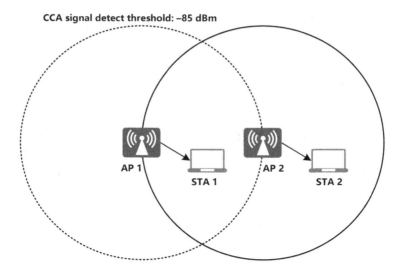

FIGURE 3.44   Scenario 1—lowered CCA signal detect threshold.

- Raise the CCA signal detect threshold for increased concurrent data transmission on co-channel APs

  As illustrated in Figure 3.45, AP 1 and AP 2 are co-channel APs, use the default CCA protocol threshold (−82 dBm), and are

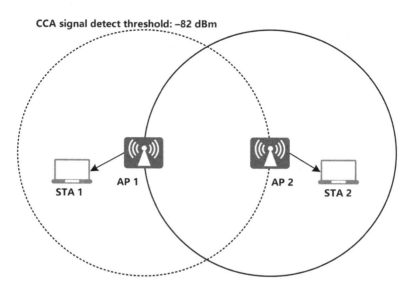

FIGURE 3.45   Scenario 2—default CCA signal detect threshold.

CCA signal detect threshold: −78 dBm

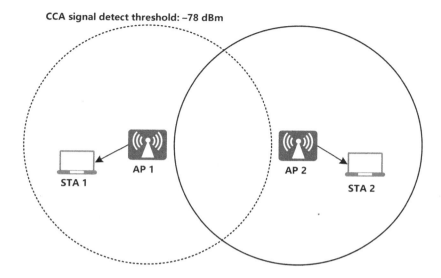

FIGURE 3.46   Scenario 2—raised CCA signal detect threshold.

aware of each other. However, STA 1 and AP 2 are unaware of each other, so are STA 2 and AP 1. When AP 1 sends packets to STA 1, AP 2 considers that the channel is busy based on the CCA mechanism, and therefore sends packets to STA 2 only after the channel becomes clear.

In this case, with STA 2 unable to sense the signals from AP 1, STA 2 can correctly parse the packets from AP 2. As illustrated in Figure 3.46, with the dynamic CCA mechanism, the scenario is identified and the CCA signal detect threshold is then raised to −78 dBm to make AP 1 and AP 2 mutually unaware. As a result, AP2-to-STA2 packet transmission and AP1-to-STA1 packet transmission will be free of mutual impact. Therefore, raising the CCA signal detect threshold in such scenarios help increase concurrent data transmission among co-channel APs, improving overall WLAN performance.

3.5.2.3 Applications of CCA

Upon detecting that channels are busy, Wi-Fi devices wait. When the channels are clear, they contend for the channels. The CCA mechanism reduces the probability of collision between interfering and desired signals and is therefore applicable to scenarios where interference is present.

Huawei's dynamic CCA mechanism adjusts the CCA signal detect threshold for APs based on scenario conditions, ensuring better performance when Wi-Fi co-channel interference is present.

3.5.3 Request To Send/Clear to Send

3.5.3.1 Technological Background

The CCA mechanism monitors channel status (clear or busy) to reduce the probability of collision. The effect of the CCA mechanism will be compromised in environments with "hidden nodes" and "exposed nodes." Therefore, the RTS/CTS mechanism is introduced.

A hidden node means that the node is within the coverage area of a receiving node, but outside that of a sending node. As illustrated in Figure 3.47, nodes A, B, and C work on the same channel. Node A is aware of node B, node B is aware of nodes A and C, node C is aware of node B, and nodes A and C are unaware of each other. In this case, node C is a hidden node "invisible" in the coverage area of node A but could potentially collide with node A. This is because, when node A is sending data to node B, node C cannot detect the transmission and considers that the channel is clear. Then, if node C also sends data to node B at this moment, the data sent by both the nodes will collide, causing parsing failures at the receiving node B.

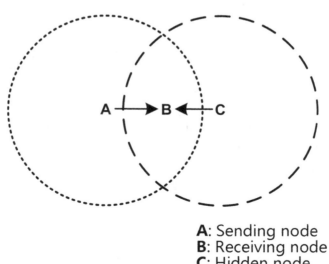

A: Sending node
B: Receiving node
C: Hidden node

FIGURE 3.47 Hidden node.

An exposed node means that the node is within the coverage area of a sending node but outside that of a receiving node. As illustrated in Figure 3.48, nodes A, B, C, and D work on the same channel. Node A is aware of node B, node B is aware of nodes A and C, node C is aware of nodes B and D, node D is aware of node C, and mutual detection is unreachable between nodes A and C, as well as between nodes A and D and between nodes B and D. In this case, when node B sends data to node A, node C detects the transmission and therefore postpones its own transmission to node D. However, the delay is unnecessary because node C is outside the coverage of node A and no collision occurs between data transmission from node C to node D as well as that from node B to node A. In this case, node C is an exposed node to node B.

In the case of hidden nodes, a contending node cannot be discovered due to excessive distances, leading to collision and undermining WLAN performance. In the case of exposed nodes, a non-contending node is undesirably detected due to its close proximity, preventing concurrent transmission from the non-contending node. 802.11 defines the RTS/CTS mechanism to address the collision issue caused by hidden nodes, and the undesired contention issue caused by exposed nodes. The mechanism introduces channel reservation (CR), allowing the sending node to reserve channels through handshakes between the sending and receiving

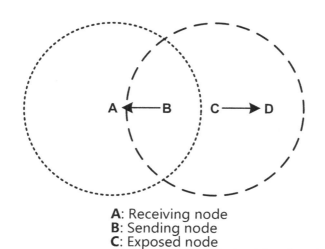

A: Receiving node
B: Sending node
C: Exposed node
D: Receiving node (waiting for data)

FIGURE 3.48　Exposed node.

nodes. As a result, a prolonged collision of data frames is avoided through a quicker exchange of RTS/CTS frames.

3.5.3.2 Working Principles
1. RTS/CTS mechanism

Figure 3.49 illustrates how data is transmitted and received with RTS/CTS.

The source node sends an RTS frame to the destination node. Collision is likely between RTS frame transmission at source nodes. With RTS frames lasting for a short time, the loss arising from the collision, if any, is limited.

After receiving the RTS frame, the destination node broadcasts the CTS frame, and other nodes that receive the CTS frame, including hidden nodes, cannot transmit any data. Then, the source node proceeds with data frame transmission and the destination node replies with an ACK frame upon receiving the entire data frame.

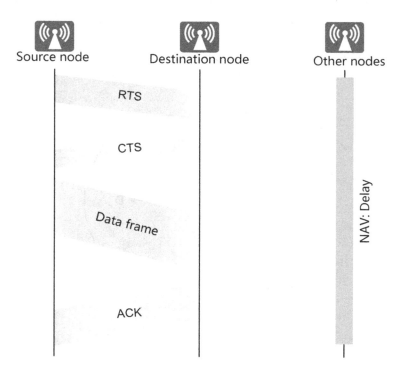

FIGURE 3.49   Data transmission with RTS/CTS.

Upon receiving the ACK frame, these nodes with their data transmission suspended will be unlocked. In addition, a node receiving the RTS frame but not receiving the CTS frame continues its data transmission without a suspension.

Though the RTS/CTS handshake mechanism is used, collision is possible. For example, if nodes A and B send RTS frames to the AP at the same, a collision will occur between the two RTS frames. This prevents the AP from correctly receiving the RTS frame, and it therefore does not send the CTS frame. As a result, nodes A and B still collide with each other, as is the case in the Ethernet. Then, they have to resend RTS frames after their respective random backoff periods have elapsed, as illustrated in Figure 3.50.

2. Huawei's dynamic RTS/CTS mechanism

The RTS/CTS mechanism reduces the impact of hidden nodes and exposed nodes on WLAN performance. With RTS/CTS, all data packet transmissions require channel reservation through RTS/CTS frame exchange. It takes time to complete RTS/CTS frame exchange, leading to extra time overhead. In radio environments in the absence of interference, hidden nodes, and exposed nodes, the RTS/CTS

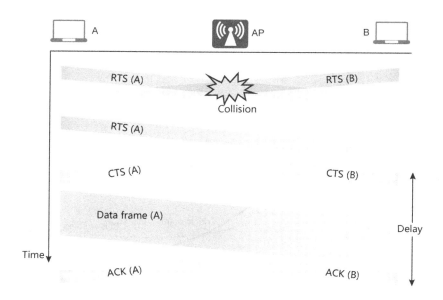

FIGURE 3.50 RTS frame collision.

mechanism works in the opposite way. This results in less time for data transmission and a decline in WLAN performance.

Huawei has introduced a dynamic RTS/CTS mechanism for WLAN products to improve the RTS/CTS performance. The enhancement includes a new process of calculating the transmission efficiency, with and without the RTS/CTS mechanism in the network, based on the collision probability. The RTS/CTS mechanism is enabled or disabled flexibly, ensuring optimal transmission efficiency.

The following two definitions apply to the dynamic RTS/CTS mechanism:

Transmission efficiency = Duration with effective data packet

transmission / Airtime occupation

Collision rate = Number of collision packets / Number of sent packet

The collision rate is determined by packet size, MCS, hidden nodes, and interference strength.

Figure 3.51 illustrates the transmission efficiency changes with the collision rate when the packet size is fixed. As the collision rate

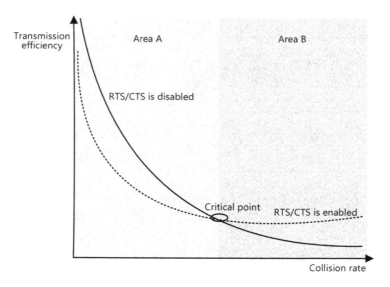

FIGURE 3.51   Transmission efficiency changes.

increases, a large number of retransmissions are required, decreasing the transmission efficiency.

The following conditions determine whether the RTS/CTS mechanism is enabled.

- When the collision rate falls into area A, where the collision rate is low, the transmission overhead caused by the RTS/CTS mechanism enabled outnumbers the performance gained. In this case, the RTS/CTS mechanism is disabled.

- As the collision rate increases, and, passing the critical point, falls into areas B, disabling the RTS/CTS mechanism will lead to a large number of packet retransmissions, significantly reducing transmission efficiency. Therefore, the RTS/CTS mechanism is enabled to effectively reduce collisions.

The dynamic RTS/CTS mechanism calculates the transmission efficiency with and without the RTS/CTS mechanism in real time based on the collision rate, packet length, MCS, and RTS/CTS frame exchange overhead. Based on the calculation results, the RTS/CTS mechanism can be enabled or disabled in the interest of higher transmission efficiency and maximized network capabilities.

In Wi-Fi 6, multiuser technology is introduced to substantially increase the number of STAs accessing networks. This leads to significantly increased overhead for RTS/CTS frame exchanges. To ensure data transmission efficiency in high-density networking, an enhanced RTS/CTS mechanism, referred to as MU-RTS/CTS, is introduced. The enhancement allows for negotiation with multiple STAs on channel reservation in a single RTS/CTS frame exchange, greatly reducing overhead.

3.5.3.3 Applications of RTS/CTS

The RTS/CTS mechanism applies to networks with a relatively high probability of hidden nodes and exposed nodes, reducing the high probability of collision caused by hidden nodes and mitigating the impact of exposed nodes on concurrent data transmissions.

How does the RTS/CTS mechanism solve the issue of hidden nodes? As illustrated in Figure 3.52, assume that node A and node C, which are mutually unaware of each other, send data to node B:

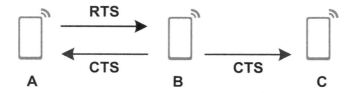

FIGURE 3.52 RTS/CTS mechanism solving the issue of hidden nodes.

1. Node A sends an RTS frame first.

2. Node C does not receive the RTS frame from node A but receives the CTS frame from node B. Therefore, node C will not send data when node A is sending data. In this case, interference for data transmission from node A to node B can be avoided.

How does the RTS/CTS mechanism solve the issue of exposed nodes? As illustrated in Figure 3.53, assume that node B sends data to node A and at the same time node C sends data to node D:

1. Node B sends an RTS frame to node A first, and both nodes A and C receive the RTS frame. After a while, however, node C does not receive the CTS frame from node A.

2. With the new mechanism, node C does not need to wait any longer because it does not receive the CTS frame from node A. This means that node A is out of the coverage of node C, and there is no interference between data transmission from node C to node B and data transmission from node B to node A.

3.5.4 AMC Algorithm

3.5.4.1 Background

Although the CCA and RTS/CTS mechanisms reduce the probability of collision caused by interference, collision still occurs occasionally. As illustrated in Figure 3.54, the RTS/CTS mechanism is enabled for all

FIGURE 3.53 RTS/CTS mechanism solving the issue of exposed nodes.

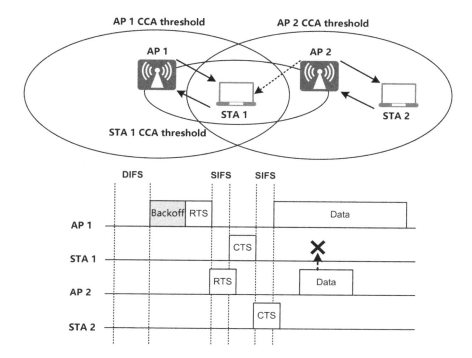

FIGURE 3.54   RTS/CTS collision.

service transmission on AP 1 and AP 2, and AP 2 is a hidden node in the downlink link between AP 1 and STA 1. Ideally, with RTS/CTS frame exchange, AP 2 can receive the CTS frame from STA 1, eliminating the impact of the hidden node. However, if AP 2 sends an RTS packet to STA 2 immediately after AP 1 sends an RTS packet to STA 1, and if it does not receive a CTS packet from STA 1, AP 2 will proceed with data transmission to STA 2. In this case, it is likely that AP 1 sends downlink packets to STA 1 while SPA 2 is sending downlink packets to STA 2, leading to collision at STA 1.

The AMC algorithm is introduced to dynamically adjust the packet-sending rates in the case of unavoidable collision to minimize the performance loss.

3.5.4.2 Principles

The AMC algorithm is one type of link adaptation technology. It dynamically adjusts MCSs based on channel conditions during packet transmission.

Signals may be subject to errors during transmission. The probability of these errors is referred to as packet error rate (PER). The PER is related to channel conditions and the MCS used for the transmission. Given the same channel conditions, a higher MCS index leads to an increased PER. Given the same MCS index, a higher channel quality allows for a decreased PER. The effective throughput of a device can be calculated based on the MCS and the PER.

$$\text{Effective throughput} = \text{Transmission rate corresponding to an MCS}$$

$$\times \text{Transmission efficiency} \times (1 - \text{PER})$$

Considering varied rates allowed by different MCSs, a smaller PER does not necessarily lead to a better effect. An excessively high PER causes excessive retransmission and even STA disconnections at worst. Conversely, an excessively low PER significantly decreases the resource utilization. The AMC algorithm aims to maintain the PER close to the threshold value through MCS adjustment. With interference present in networks, the PER changes as the level of interference changes. The AMC algorithm adjusts the MCS in response to the PER changes, ensuring optimal effective throughput achievable on APs.

Figure 3.55 illustrates the basic framework of AMC implementation for wireless devices:

- The software layer constructs packets and adds them in the sending queue.

- The software layer selects the optimal MCS based on current channel status and negotiation results.

- The hardware uses the MCS provided by the software layer to send the packets and feeds back the sending result to the software layer.

- The software layer updates the PER information based on the hardware feedback.

In the presence of severe Wi-Fi co-channel interference, the traditional AMC algorithm causes the MCS indexes to decrease continuously. At worst, WLAN services will be unavailable. This is because the AMC algorithm puts a cap on the maximum rate (rateMaxPhy) used for rate control,

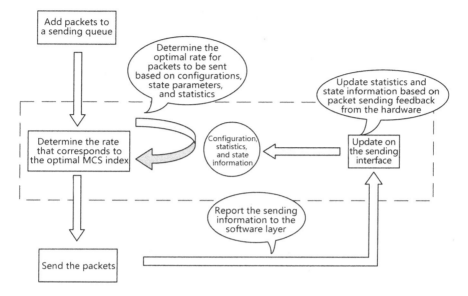

FIGURE 3.55   Basic framework of AMC implementation.

so as to limit the speed at which the MCS index increases after the interference is cleared. rateMaxPhy is configured to prevent a high-order MCS from being used during rate control before channel measurement, avoiding incorrect rate selection caused by an incorrect PER estimate. rateMaxPhy can be adjusted based on the statistics of PER.

- Conditions for rateMaxPhy decrease: The PER that corresponds to the current rate is above the PER threshold and the rate is smaller than the rateMaxPhy value.

- Conditions for rateMaxPhy increase: N consecutive physical layer convergence procedure Protocol Data Units (PPDUs) are successfully transmitted at the current rate.

In the presence of severe Wi-Fi co-channel interference, if the interference leads to the PER being above the threshold of rateMaxPhy decrease, the rateMaxPhy value will decrease continuously. Due to Wi-Fi co-channel interference, the conditions for rateMaxPhy increase can hardly be satisfied. As a result, the rateMaxPhy value will only decrease with no chance for an increase, causing the lowest transmission rate to be used. In the meantime, the SNR or RSSI is not low, and the rate decrease further prolongs the packet transmission time, increasing the probability of collision. Therefore, rate reduction cannot address Wi-Fi co-channel interference.

Huawei improves the AMC algorithm with rate protection for WLAN products to address the issue of continuous rateMaxPhy decrease and declined transmission rates caused by Wi-Fi co-channel interference. Wi-Fi networks are Time Division Duplex (TDD) based, and the uplink SNR/RSSI value represents the channel quality to some extent. If the SNR/RSSI value is large, the radio link quality is good and it is unreasonable to select a smaller MCS index. Therefore, a protected rate is needed in the original AMC algorithm to prevent the transmission rate from continuous decrease. The rate protection algorithm associates an uplink SNR/RSSI value with a $MCS_{protection}$ value to execute protection on the AMC algorithm. This helps prevent the MCS index from continuous decrease in networks suffering severe interference.

As illustrated in Figure 3.56, prior to the AMC algorithm improvement, the rateMaxPhy value continuously decreases without a chance of increase, causing the smallest MCS index to be used. As a result, the transmission efficiency sharply decreases and, in the worst cases, WLAN services will be interrupted. With the AMC algorithm improvement, $MCS_{protection}$ is used, preventing MCS indexes from becoming smaller than $MCS_{protection}$ in the presence of good signal quality. This ensures transmission efficiency and improves user experience.

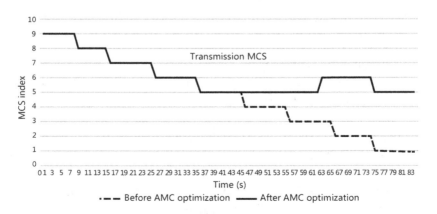

FIGURE 3.56   AMC optimization effect.

3.5.4.3 Applications

In the event of collision, the AMC algorithm adjusts MCS indexes for sending packets, minimizing the impact of collision on network performance, and ensuring optimal performance on the network. Therefore, the AMC algorithm is applicable to the scenarios, especially in highly dense networks, where strong interference is present.

Take stadiums for example. APs are densely deployed, causing strong co-channel interference and leading to a high rate of collision. The AMC algorithm calculates the effective throughput with different MCS indexes based on the transmission rates, transmission efficiencies, and PERs corresponding to these MCSs. It then selects the MCS index that ensures the highest effective throughput. As a result, the performance loss caused by collision is reduced, improving user experience.

3.6 AIR INTERFACE QOS

The preceding sections describe methods for optimizing air interface performance at the physical and access layers. These methods are not service specific and apply the same priority to all services. In enterprise and home WLANs, voice, video, and gaming services require high real-time performance and network stability, as delay and frame freezing will lead to significant deterioration of user experience. It is obviously not practical to apply the same priority to these services as that assigned to web browsing, email, and other common services. This section describes how to ensure WLAN QoS at the application layer.

QoS is a service that ensures the end-to-end requirements of network services, including bandwidth, delay, jitter, and packet loss. It aims to provide dedicated service quality to various applications. In particular, it allows networks to provide sensitive services with guaranteed high bandwidth and low delay without impacting common service experiences. In WLANs, QoS represents differentiated services provided to meet diversified traffic requirements of various users. As illustrated in Figure 3.57, WLAN QoS is further classified into wireless QoS (AP-STA) and wired QoS (AP-WAC/switch).

- Wireless QoS: Based on 802.11e –Wi-Fi Multimedia (WMM), enhanced scheduling algorithms provided by vendors, such as dynamic EDCA and airtime, are implemented to complete radio resource management and congestion control specific to user and service types.

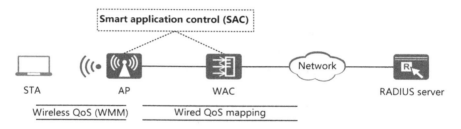

FIGURE 3.57 Wireless QoS and wired QoS.

- Wired QoS: Voice and video traffic feature matching is implemented to identify packet types, and QoS mapping is performed for wired services based on IEEE 802.1p and other standards to provide traffic priority scheduling and other services.

3.6.1 Dynamic EDCA

3.6.1.1 Technological Background

Having served as the standard since a long time, 802.11e can no longer ensure optimal WLAN QoS requirements. As a result, delay-sensitive services such as those involving voice and video are unable to deliver satisfactory user experiences. To address this issue, the Wi-Fi Alliance initiated the establishment of the WMM standard.

Put simply, dynamic EDCA is introduced to ensure the sequence of speech and prioritize the delivery of important content.

3.6.1.2 Principles of Huawei's Dynamic EDCA

1. Introduction to EDCA

Prioritized delivery of important content requires content importance to be assessed and granted a higher priority that corresponds to the level of importance. Based on 802.11, if a collision occurs when multiple nodes attempt to transmit data, random numbers are selected to queue data transmission. To ensure preferential transmission of high-priority content for each node, a smaller random number is assigned to higher-priority content while a larger random number to lower-priority content. As a result, the nodes with the smaller random number have a higher chance of transmission. In addition, by reducing the wait time, more opportunities for transmission become available. Furthermore, a longer transmission time window is allocated to higher-priority content, while the shortest possible window is assigned to lower-priority content.

To summarize, the WMM standard divides the EDCA mechanism defined in 802.11e into sorting service priorities, and, for higher-priority services, lowering the value ranges of the random queue number, reducing the wait time for transmission, and increasing the sending duration.

In descending order of priority, EDCA divides data packets into four access category (AC) queues: AC_VO (Voice), AC_VI (Video), AC_BE (Best Effort), and AC_BK (Background). A higher-priority AC queue has a higher probability of using a channel.

Figure 3.58 illustrates how an AC queue contends for channels. Table 3.3 provides the mapping between the four AC queues and user preferences (UPs) in 802.11. A larger UP value indicates a higher priority.

Each AC queue corresponds to a set of EDCA parameter configurations, in order to determine its probability of channel usage. This aims to ensure that a higher-priority AC queue has a higher probability of channel usage than a lower-priority AC queue. Figure 3.59

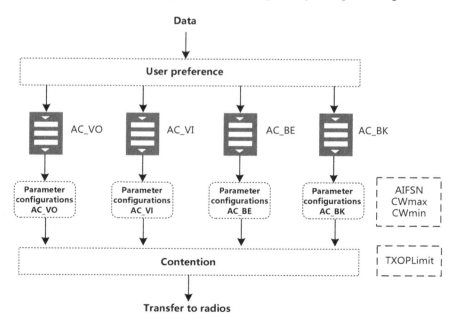

FIGURE 3.58   Channel contending of AC queues.

TABLE 3.3   Mapping between AC Queues and Ups

UP	AC
7	AC_VO (Voice)
6	
5	AC_VI (Video)
4	
3	AC_BE (Best Effort)
0	
2	AC_BK (Background)
1	

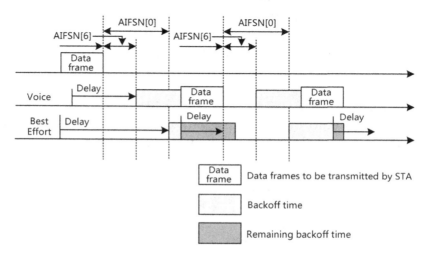

FIGURE 3.59   Principles of EDCA channel contention.

illustrates the principles of EDCA channel contention. In addition to having a shorter backoff time, voice packets have a smaller arbitration inter-frame spacing number (which is AIFSN[6]) than Best Effort packets (which is AIFSN[0]). When both voice and BE packets need to be sent, voice packets will preferentially use a radio channel because they have a higher UP.

Table 3.4 describes the parameters defined in EDCA. The set of parameter settings is specific to AC queues.

As mentioned above, content priority is classified and sorted. A providing node will first assign the content a priority based on urgency before transferring it to the sending node. After receiving the content, the sending node will also assign the content a priority.

TABLE 3.4   Parameters Defined in EDCA

Name	Description
AIFSN	In the distributed coordination function (DCF) mechanism, the DCF inter.frame space (DIFS) is fixed, and the WMM uses different DIFS values based on ACs. A larger AIFSN value means a longer waiting time and a lower priority
ECWmin and ECWmax	The two parameters together determine the average backoff time, with greater values indicating a longer average backoff time and a lower priority
TXOPLimit	The parameter determines the maximum time permitted to use a successfully contended channel, with a larger value indicating a longer channel use time. If the value is **0**, only one data frame can be sent by a winning contender using the channel

If the priorities are inconsistent, the sending node will map between these assigned priorities.

Priorities differ on wireless and wired networks. At the wireless network, UP is carried in packets to indicate the priority. At the wired VLAN, 802.1p priorities are used, while at the Internet Protocol (IP) side, differentiated services code point (DSCP) priorities are used. Therefore, priority mapping must be configured to maintain packet priority.

As illustrated in Figure 3.60, in the uplink, after receiving an 802.11 packet from an STA, the AP maps the UP to the 802.1p or DSCP priority used at the wired side. In the downlink, after receiving an 802.3 packet, the AP maps the DSCP or 802.1p priority to the 802.11 UP.

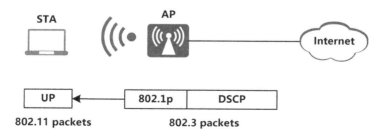

FIGURE 3.60   Priority mapping.

2. DSCP priority field

In an IP packet header, bits 0–5 in the Type of Service (ToS) field are used to indicate a DSCP, as illustrated in Figure 3.61. With a total of six bits used, a total of 64 priorities (0–63) can be defined.

3. 802.1p field

As defined in IEEE 802.1Q, the PRI field (802.1p priority) in the Ethernet frame header identifies the QoS requirements between two layer 2 devices. Figure 3.62 illustrates the PRI field, or referred to as Class of Service (CoS), in Ethernet frames.

The 802.1Q PRI field in the Ethernet frame header contains three bits, defining a total of eight priorities: 7, 6, 5, 4, 3, 2, 1, and 0 in descending order.

These priorities are used to classify and recognize packet priorities on the wired networks. To transmit packets between STAs and the AP/WAC, priority mapping is required to ensure QoS processing for packets between the air interface and the wired networks. Table 3.5 provides the default mappings from the DSCP priorities to 802.1p (CoS) and 802.11e UPs (WMM).

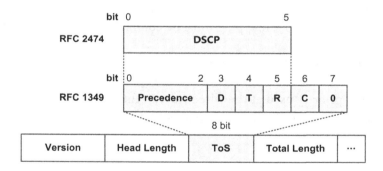

FIGURE 3.61   DSCP priorities of IP packets.

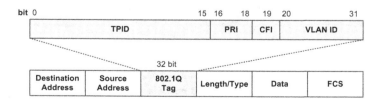

FIGURE 3.62   802.1p priorities of VLAN frames.

TABLE 3.5   Default Mappings from the DSCP
Priorities to 802.1p (CoS) and 802.11e UPs (WMM)

DSCP	802.1p (CoS)	802.11e UP (WMM)
0–7	0	0
8–15	1	1
16–23	2	2
24–31	3	3
32–39	4	4
40–47	5	5
48–55	6	6
56–63	7	7

4. Huawei's dynamic EDCA

Common EDCA parameters are statically configured. Under the demands of complex services, various user types, and changing data loads, such static configuration cannot meet the necessary requirements. This leads to collision as a result of small contention windows or a waste of resources due to large contention windows, as illustrated in Figure 3.63.

Huawei further enhances standard EDCA based on the traffic characteristics of application-layer services. Contention windows for the uplink and downlink transmissions are dynamically adjusted based on factors such as STA quantity, collision probability, channel usage,

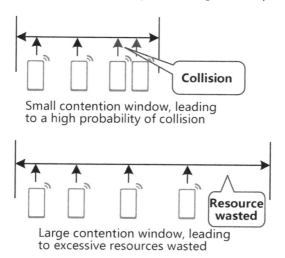

FIGURE 3.63   Static EDCA parameter settings unable to meet requirements.

and real-time load. This allows EDCA parameter settings to be more adaptive so that the configurations are optimal at all times. Collisions caused by random backoff and small contention windows are minimized, and channel resource waste due to large contention windows is avoided. As a result, throughput can be ensured in multiuser networks. Figure 3.64 illustrates the channel contention with dynamic EDCA.

Dynamic EDCA dynamically adjusts EDCA parameter configurations based on historical and current loads. Figure 3.65 illustrates the process of load-based EDCA parameter adaptation.

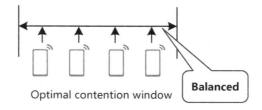

FIGURE 3.64   Channel contention with dynamic EDCA.

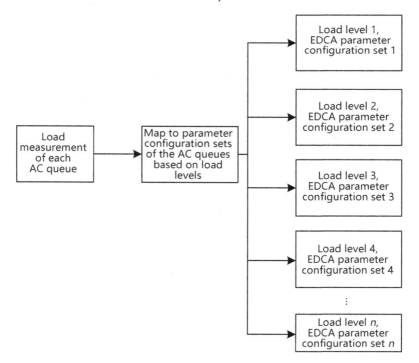

FIGURE 3.65   Process and parameter settings of load-based EDCA.

Load-based EDCA parameter adaptation includes the following:

- AC queue load measurement: Measures the air interface transmission duration of data packets to be sent in each AC queue to calculate the load to be scheduled.

- Parameter configuration policy selection: Selects a set of parameter configurations based on the current load proportion of the AC queue.

3.6.1.3 Applications and Benefits of Dynamic EDCA

Traffic-hungry and delay-sensitive wireless voice and video services are common in enterprise mobile office scenarios. With dynamic EDCA enabled, voice and video service loads are measured and their proportions calculated. EDCA parameters are configured more accurately based on the results, further reducing the probability of collision and improving overall throughput. This technique delivers an effective boost to the user experience of mobile office voice and video services.

3.6.2 Voice and Video Packet Identification

3.6.2.1 Technological Background

While voice and video services demand a high degree of real-time performance and reliability, such services are often not tagged for high priority on live networks. In the absence of identification, they are treated as common services and mapped onto lower priorities, putting service quality at risk. As such, effective identification is required for video and voice packets so that accurate QoS policy control can be implemented.

3.6.2.2 Technological Principles

Feature codes are created from extracted packet information, which includes specific IP addresses, domain name systems, traffic, ports, character strings, sequences, and sequence orders. With enough packet feature codes, feature modeling or training can be completed, resulting in a feature library which enables data packets to be identified by checking their feature codes against those already indexed. This feature library is updated either in real time or periodically based on service changes. During identification, multiple groups of packets must be collected, analyzed, and checked against the feature library to identify the packet type, particularly for voice and video services of high importance. Based on the identification results, specialized QoS policies can be implemented, or custom optimization performed, for voice and video applications, to improve service quality.

Smart application control (SAC) is an intelligent application protocol identification and classification engine developed by Huawei for WLAN products, and is distributed on both APs and WACs in networks. SAC detects and identifies layers 4–7 content of data packets and content of dynamic protocols, such as a Hypertext Transfer Protocol (HTTP) and a Real-time Transport Protocol (RTP). Figure 3.66 illustrates how SAC works.

Application protocol packets are identified based on their feature codes. However, as application programs often receive regular updates, their associated packet feature codes can change. To ensure the effectiveness of this identification approach, feature codes need to be updated regularly so that they can accurately match the application protocol. In cases where feature codes are configured within software packages, the software will also need to be updated, and this can significantly affect services. If feature library files are separated from software packages, as already practiced by some vendors, feature code files can be loaded and updated at any time without affecting other services. With enough analysis of common applications, feature library files can be generated and loaded to devices in predefined mode, enabling the update of predefined feature library files.

After the SAC traffic statistic function is enabled, the device automatically identifies and measures traffic from various applications. This enables network administrators to understand network traffic in real time and take appropriate measures to optimize deployment and adjust bandwidth allocation.

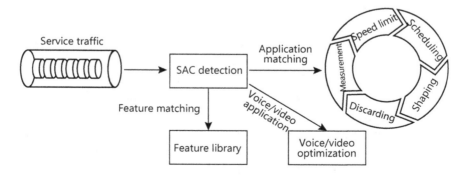

FIGURE 3.66   Working mechanism of SAC.

3.6.2.3 Applications and Benefits

Huawei's SAC feature library can efficiently identify most voice and video services, and is particularly well suited to the identification of the following:

- Session Initiation Protocol (SIP) voice and video services
- RTP voice service
- Microsoft Lync and Skype for Business
- Tencent QQ
- Tencent WeChat
- WeLink on-demand video services
- DingTalk

The above voice and video services cover most enterprise and home scenarios, and identifying these services will significantly improve user experience.

3.6.3 Airtime Scheduling

3.6.3.1 Technological Background

EDCA ensures that higher-priority content is prioritized, and that nodes are transmitting this type of content in a fair manner. Intermittent transmission as a result of interference or long distance requires repeated retransmissions, decreasing overall efficiency while reducing the time available for transmission by other nodes. As such, the duration of intermittent transmission must be minimized to ensure fairness.

In WLANs, as illustrated in Figure 3.67, STA-perceived data rates vary greatly as a result of fluctuating performance levels or radio environments. Given the same size of packets, low-speed STAs require a longer time to transmit over the air interface than high-speed STAs. This adversely affects other STAs under the AP (especially high-speed STAs) and decreases the overall throughput of the AP.

To minimize the impact, the AP performs access control over low-speed STAs during the access phase based on signal strength. However, this method invites unfairness as it restrains low-speed STAs from accessing networks. At the network side, it is also challenging to limit low-speed

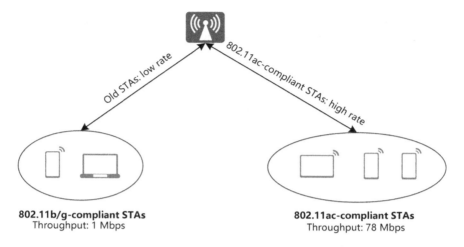

802.11b/g-compliant STAs
Throughput: 1 Mbps

802.11ac-compliant STAs
Throughput: 78 Mbps

FIGURE 3.67   User experience compromised as a result of low-speed STAs.

STA access requests, leading to such requests being repeatedly sent. As a result, extra air interface overhead is required.

Such problems highlight the need for additional methods capable of providing solutions. As downlink traffic still accounts for the majority of current network data flows, time-based fair scheduling is an ideal option to ensure fairness, improve user experience, and increase the downlink throughput of the AP during downlink traffic scheduling. This approach also avoids an extra load on the air interface caused by repeated low-speed STA access requests. By ensuring network access for both high-speed and low-speed STAs, a significant improvement of both channel usage fairness among WLAN STAs and overall user experience can be achieved.

3.6.3.2 Technological Principles

Based on the WMM-defined EDCA mechanism, Huawei's airtime scheduling implements time-based fair scheduling among STAs performing the same type of services. The scheduled STAs then obtain EDCA settings based on AC queues and contend for air interface resources, as illustrated in Figure 3.68.

Airtime switches from first-in-first-out (FIFO) to time-based scheduling of AC queues in WMM, helping to ensure an equal accumulated scheduling time among STAs of different rate capabilities. Before sending packets, the AP selects the STA with the minimum accumulated scheduling time

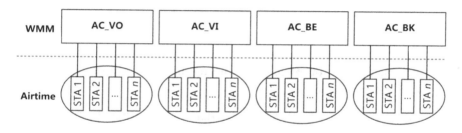

FIGURE 3.68   Relationship between airtime and WMM scheduling.

to transmit first, ensuring approximate air interface resources (in terms of time) for users in a queue. As illustrated in Figure 3.69, STAs of different rate capabilities in the AC queue are assigned the same scheduling time when airtime scheduling is used. This prevents low-speed STAs from using the air interface for an excessively long time, while also providing high-speed STAs with more opportunities for data transmission.

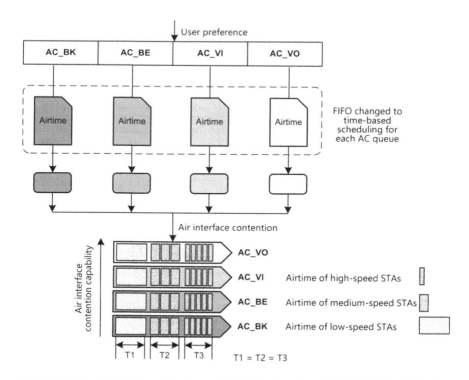

FIGURE 3.69   Principles of airtime scheduling based on WMM-defined EDCA.

Airtime scheduling works as follows:

- Inserts a new STA to a position determined according to radio channel use time, as opposed to traditional methods of placing it at the end of the user queue.

- Checks whether the STA still has data to send after the first queue is scheduled. If it does, it is added into the AC queue based on radio channel use time, and the STA having the shortest channel use time is scheduled. Otherwise, the next user is directly scheduled.

As illustrated in Figure 3.70, there are four STAs requesting data transmission under the same radio. For STA 1, the current channel use time is 3 and the time needed to transmit its data is 2. For STA 2, these time values are 4 and 4, respectively. For STA 3, the values are 6 and 6, and for STA 4, the values are 7 and 7.

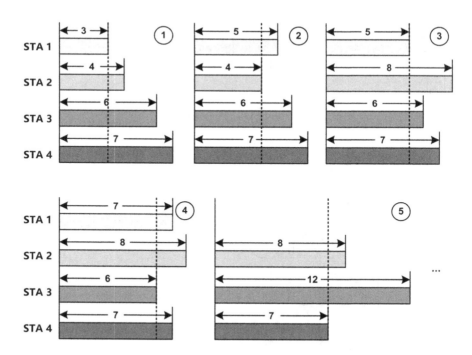

FIGURE 3.70 Airtime scheduling process.

After airtime scheduling is enabled, and based on the current channel use time of the four STAs (3, 4, 6, and 7), STA 1 has the shortest channel use time and therefore is scheduled first.

Once STA 1 finishes data transmission, 2 is added to its channel use time, bringing its current value to 5. The channel use times for each of the four STAs are now 5, 4, 6, and 7, respectively. As STA 2 has the shortest channel use time, it is preferentially scheduled.

After STA 2 finishes data transmission, 4 is added to its channel use time, bringing its current value to 8. The channel use times for each of the four STAs are now 5, 8, 6, and 7, respectively. As STA 1 has the shortest channel use time, it is preferentially scheduled.

After STA 1 finishes data transmission, and with no more data to transmit, channel use time is sequenced among the remaining three STAs, which are 8, 6, and 7, respectively. STA 3 has the shortest channel use time and is preferentially scheduled.

After STA 3 finishes data transmission, 6 is added to its channel use time, bringing its current value to 12. The channel use times for the remaining three STAs are now 8, 12, and 7, respectively. As STA 4 has the shortest channel use time, it is preferentially scheduled.

In each data transmission, the STA with the shortest accumulated channel use time will be preferentially scheduled, ensuring fair radio channel usage.

The radio channel use time of all devices is periodically cleared to avoid first access users being unable to utilize radio channels and ensure an equal channel use weight among all STAs.

3.6.3.3 Applications and Benefits

In certain outdoor scenarios, STAs are sometimes located close to the AP and are subject to reduced path loss, receiving strong signals as a result. In contrast, distant STAs experience large path loss which leads to reduced receiving signal strength.

As illustrated in Figure 3.71, STAs located close to the AP receive stronger signals and experience higher throughput than those positioned further away. In this case, airtime scheduling prevents low-speed STAs from using air interfaces for an excessively long time, improving user experience of high-speed STAs while also increasing overall AP throughput.

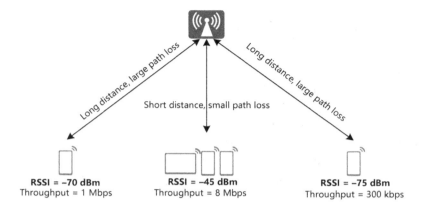

FIGURE 3.71 Airtime scheduling mechanism.

3.7 ANTENNA TECHNOLOGIES

In wireless systems, antennas serve as the interface between transceivers and signal propagation media. Their functions include receiving and sending radio waves. In WLANs, radio signals sent by transmitters are propagated in the form of electromagnetic waves and received by receive antennas. Therefore, antennas and advanced antenna technologies are essential for maximizing signal coverage and optimizing WLAN performance.

This section describes WLAN antennas and Huawei's advanced antenna technologies in terms of types, main specifications, smart antenna technologies, and high-density antenna technologies.

3.7.1 Types and Main Specifications

3.7.1.1 Antenna Types

Antennas are a category of devices used to send or receive radio waves in broadcast, television, point-to-point radio communications, radar, space exploration, and many other systems. Antennas work in the air, in outer space, underwater, or even in soil and rocks. In the physical science, an antenna is the combination of one or more conductors that produce electromagnetic fields as a result of applied time-varying voltages or currents, or yield time-varying currents as a result of induction in electromagnetic fields and provide time-varying voltages at the output end.

Antennas assign wireless systems three basic properties: gain, directivity, and polarization. Specifically, gain is a measure of energy amplification, directivity measures the degree to which the radiation emitted is concentrated in a single direction, and polarization is the direction of the electric field of the radio wave. Though antennas radiate energy in all directions in free space (FS), when with specific structures, their radiation is significant in one direction while ignorable in others. Based on certain three dimensional (regularly vertical or horizontal) planes, antennas are divided into three types, as illustrated in Figure 3.72.

- Omnidirectional antenna: The electromagnetic energy is radiated equally well in all horizontal directions but differently in all vertical directions. Its antenna pattern is similar to the light from filament lamps as it spreads in all directions on the horizontal plane.

- Directional antenna: The electromagnetic energy is radiated differently in all horizontal and vertical directions. Its antenna pattern is similar to the light from torches as it spreads toward certain directions. Given the same electromagnetic energy, although directional antennas extend radio coverage further in one direction, they lose the ability to provide coverage in other directions.

- Smart antenna: Horizontally, multiple directional radiation modes and one omnidirectional radiation mode are provided, while signals from terminals are received in omnidirectional radiation mode. Smart antenna algorithms are a directional radiation mode in which

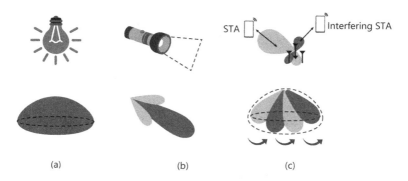

(a)　　　　　　(b)　　　　　　(c)

FIGURE 3.72 Antenna categories based on horizontal beam characteristics. (a) omnidirectional antenna, (b) directional antenna, and (c) smart antenna.

the locations of terminals are determined according to the signals received from the terminals and the Central Processing Unit is then controlled to direct the maximum amount of radiation toward the STAs based on the locations.

3.7.1.2 Main Specifications

An antenna's performance is determined by many specifications, including resonance frequency, impedance, gain, aperture or radiation pattern, polarization, efficiency, and bandwidth. These specifications can all be adjusted during the antenna design process. The following defines the abovementioned antenna specifications:

1. Resonance frequency

   For an antenna, the resonance frequency is related to its electrical length (the ratio of the physical length of the transmission line to the wavelength of an electromagnetic wave traveling through the line). The antenna is generally tuned at a frequency and is effective on a segment of band centered on the resonance frequency. With other antenna specifications (especially radiation pattern and impedance) varying with frequencies, the resonance frequency is close only to the center frequency involved in these specifications.

2. Bandwidth

   The bandwidth of an antenna is the effective range of operating frequencies it works at, and it is usually centered on the resonance frequency. A number of approaches can be used to increase antenna bandwidth, for example, using thick metal wires, replacing metal wires with metal cage, using the components that have thinner tips (such as feeder horns), and integrating multiple antennas into a single component (characteristic impedance is used to select the antennas). Small antennas, although they are convenient to use, have inevitable limitations in bandwidth, size, and efficiency.

3. Impedance, SWR, and return loss

   Impedance is similar to the refractive index in optics. Electromagnetic waves are exposed to varied impedance when passing through various parts of antenna systems (such as radio stations, feeders, antennas, and free space (FS)). At each port, due to impedance matching, a portion of energy is reflected back to the source, forming a standing wave on the feeders. The resulting reflection loss is referred to as return loss, while the ratio of the maximum energy to the minimum energy is referred to as standing wave ratio (SWR), which is 1:1 ideally. A 1.5:1 SWR is considered a critical value in low-power consumption applications that are more sensitive to power consumption. An up to 6:1 SWR can be found in specific devices. A decreased impedance difference (impedance matching) at each port leads to a reduced SWR and improved energy transmission efficiency for antenna systems.

A high SWR will decrease power output and shorten communication distance. In WLANs, the SWR must be less than 2:1. The SWR also determines the operating bands of antennas. As it is related to the strength of reflected signals, the SWR is likely to increase if metal obstacles are close to the antenna's main lobe due to the reflection of electromagnetic waves on the metal obstacles.

4. Port isolation

Port isolation, measured in units of dB, is the ratio of the power of a local oscillator or radio signal leaking to other ports to the input power. The measures taken to increase isolation include increasing physical space, using metal isolation strips, suppressing ground currents, and adopting cross polarization, high impedance surfaces, and wave absorbers. The isolation between WLAN antenna ports must be less than 20 dB.

5. Envelope correlation coefficient (ECC)

Multiple-input multiple-output (MIMO) is based on space-time signal processing. On the basis of the time dimension, the space dimension is added through multiple independent fading copies of signals transmitted by multiple antennas to implement multidimensional signal processing that produces spatial multiplexing gains. Under the circumstance, the spatial correlation of the antenna system will reduce their spatial degree of freedom and causes channel capacity to decrease. Given the Rayleigh channel conditions, the envelope correlation coefficient ρ_e is defined using the formula below:

$$\rho_e = \frac{\left| \iint_{4\pi} \left[F_1(\theta,\phi) \cdot F_2^*(\theta,\phi) \right] d\Omega \right|^2}{\iint_{4\pi} \left| F_1(\theta,\phi) \right|^2 d\Omega \iint_{4\pi} \left| F_2(\theta,\phi) \right|^2 d\Omega}$$

6. Gain

In antenna design, gain (G) is defined as the logarithmic value of the ratio of the maximum radiation intensity produced by an antenna to that provided by a reference antenna.

$$G = 10\lg(E \cdot D)$$

In this formula, E represents the antenna efficiency, indicating how much electrical energy is converted into electromagnetic energy. D indicates the antenna's directivity.

$$D = 4\pi\left(\frac{U_{\max}}{P_{\text{rad}}}\right)$$

In this formula, $U_{\max}$ represents the transmit power in the direction of the maximum transmit power, and P_{rad} represents the total transmit power in all directions.

$$P_{\text{rad}} = \int_0^{2\pi}\int_0^{\pi} U(\theta,\phi)\sin(\theta)d\theta\,d\phi$$

If the reference antenna is an omnidirectional antenna, gain is measured in units of dBi. Considering that electromagnetic waves in a vacuum are dispersionless transverse waves with a photon rest mass of 0 and a spin of 1, it is impossible to have ideal omnidirectional reference antennas with spherical symmetry. Therefore, dipole antennas are used as reference antennas, and the antenna gain is in dBd. If the gain of an antenna is 16 dBd, for example, the gain would be 18.14 if expressed in dBi, when the reference dipole antenna has a gain of 2.14 dBi.

Antenna gain is a passive indicator, meaning that antennas do not increase power to electromagnetic waves. Instead, they redistribute the electromagnetic power with more energy radiated in one direction than in others when compared to omnidirectional antennas. Positive gains in some directions will inevitably lead to negative gains in other directions, as governed by the law of conservation of energy. Therefore, a balance must be achieved between antenna coverage and antenna gain. One example would be the parabolic antennas equipped on spacecraft to enable a high gain but a narrow coverage area so as to maintain its direction to the Earth. By contrast, the antennas used in broadcast systems provide a wide coverage area, limiting them to a small gain.

For parabolic antennas, the gain they achieve is proportional to their aperture (reflection zone), reflection surface accuracy, and transmit/receive frequencies. In general, a larger aperture and a higher frequency both lead to a higher gain. However, on a high frequency, a surface accuracy error may significantly decrease the gain.

The apertures and radiation patterns are closely related to antenna gains. An aperture is a two-dimensional beam cross-section in the direction of the maximum gain (sometimes approximately represented by the radius of the circle on the cross-section or the angle of the beam cone). A radiation pattern is a three-dimensional graph used to express antenna gains. Generally, only the horizontal and vertical two-dimensional cross-sections in the radiation direction are considered. There are usually minor lobes in the radiation patterns of high-gain antennas. A minor lobe is one of the beams other than·the main lobe (the beam with the highest gain), and it includes side lobes and back lobes. In radar systems and other systems where directions need to be determined, minor lobes adversely impact signal quality and decrease the power of the main lobe. Figure 3.73 illustrates an antenna radiation pattern.

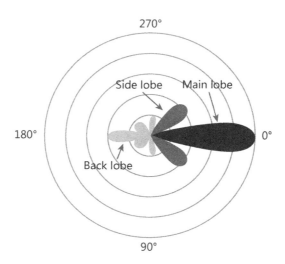

FIGURE 3.73 Main lobe, side lobe, and back lobe of an antenna radiation pattern.

The gains of typical antennas in WLANs are 2–3 dBi for external rod antennas used indoors, 3–5 dBi for internal antennas used indoors, 6–8 dBi for external omnidirectional antennas used outdoors, and 8–14 dBi for internal sector/directional antennas used outdoors.

7. Half-power beamwidth

Half-power beamwidth is the beamwidth between two points whose power is half of (specifically, 3 dB lower than) the power in the direction with the strongest radiation on the horizontal or vertical plane. It can be a horizontal or vertical beamwidth. The typical horizontal and vertical half-power beamwidths of WLAN antennas are as follows. For indoor (ceiling mounted) omnidirectional antennas, the horizontal and vertical values are 360° and 120°–150°, respectively. For outdoor (pole mounted) sector antennas, the horizontal and vertical values are 60°–90° and 10°–30°, respectively. For antennas in high-density scenarios (stadiums), the horizontal and vertical values are 30°–60° and 30°–60°, respectively.

8. Minor level

The minor level is the ratio of the maximum side lobe level to the maximum main lobe level. To increase frequency reuse efficiency and minimize co-channel interference, side lobes directing interference areas must be reduced as much as possible during AP beamforming to increase the ratio of desired signals to undesired signals. The typical minor level of WLAN antennas ranges from −16 dB to −10 dB.

9. Front-to-rear ratio

The front-to-rear ratio is the ratio of the power density in the forward maximum radiation to the power density in the backward maximum radiation within ±30°, as illustrated in Figure 3.74. It is valid only for directional antennas and indicates an antenna's capability to suppress backward interference. In WLANs, the typical front-to-rear ratio of outdoor sector antennas is greater than 20 dB.

10. Polarization

Polarization is the direction of the electric field of the radio wave. Unless otherwise specified, the direction of the electric field in the maximum radiation direction of an antenna is taken as the polarization.

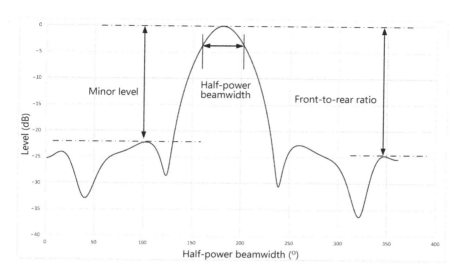

FIGURE 3.74   Half-power beamwidth of antenna lobes.

A linearly polarized wave is the electromagnetic wave in which the electric field vector orientation in space does not change. Using the ground as an object of reference, horizontally polarized waves are the waves with electric field vector orientation parallel with the ground, while vertically polarized waves are the waves with electric field vector orientation vertical to the ground. The electric field vector orientation in space is not always fixed. If the track of the vector point of the electric field is a circle, the wave is a circularly polarized wave. If the track is an ellipse, the wave is an elliptically polarized wave. Both circularly and elliptically polarized waves have the characteristics of phase rotation.

The band for electromagnetic propagation determines the polarization mode to be used. That is, in mobile communications, vertical polarization is most commonly used, whereas in broadcast systems, horizontal polarization is typically used, and elliptical polarization is most commonly used for satellite communications.

Antennas mainly include dipole antennas and microstrip slot antennas. The polarization direction of dipole antennas is the same as the direction of the dipole axis. The polarization direction of microstrip slot antennas is the same as the length direction of the slot. Therefore, the polarization direction of an antenna can be easily discriminated.

In WLANs, antennas are divided into single polarization antennas and dual polarization antennas, which are different in linear polarization modes. Dual polarization antennas use polarization diversity to reduce the detrimental effects of multipath fading so as to improve the quality of received signals for APs. Generally, dual polarization means that radiation is polarized both horizontally and vertically.

WLAN antennas feature complicated horizontal and vertical polarization. Unlike base station antennas, the polarization of internal antennas of settled APs used indoors is equivalent to the radiation field polarization of radial or circular currents in cylindrical coordinate systems, rather than simply to that of line currents. For on-PCB antennas or planar inverted F antennas (PIFAs), linearly polarized waves do not display uniform patterns in all directions. In a similar way to base stations, the external antennas of WLAN APs use either 0°/90° or +45°/−45° orthogonal linear polarization. Chapter 9 will further detail antenna selection in various scenarios.

3.7.2 Smart Antenna Technologies

The continued development of the 802.11 standards drives a significant increase in physical-layer rates on Wi-Fi networks. Mainstream products that support 802.11ac Wave 2 support a maximum of four spatial streams and up to 160 MHz bundled bandwidth, increasing the maximum transmission rate to 3.47 Gbps, a few dozen times that delivered with 802.11a, 802.11g, and 802.11n. Products that are compliant with the 802.11ax standards support an even higher maximum transmission rate of 10 Gbps, several times greater than 802.11ac Wave 2.

However, increased physical-layer rates cannot be delivered without support from radio environments. In networks, complex radio environments lead to weak coverage or coverage holes. Even in areas where signal coverage is already good, higher-quality signals are required to support higher transmission rates.

To further improve radio signal quality, mainstream vendors are introducing smart antenna technology already widely used in 4G and other public mobile networks to Wi-Fi networks. By doing this, Wi-Fi signal coverage issues will be addressed through technologies used for public mobile networks, while Wi-Fi network capacity and user experience will also be improved.

The signal coverage of AP edge users is currently also a major bottleneck of Wi-Fi networks. As most APs use omnidirectional antennas, their limited antenna gains restrict signal coverage to only those users who are located close to APs. In addition, users in the middle or at the edge of AP radio coverage areas can only access low-rate services or even have no service access, while achieving high throughput for users behind obstacles, typically poles and walls, is another big challenge for APs. In high-density networking, multiuser concurrency greatly increases inter-link interference. Working with downlink multiuser multiple-input multiple-output (MU-MIMO) introduced in 802.11ac to further improve downlink throughput and transmission efficiency is a key challenge of smart antennas to enhance AP competitiveness.

3.7.2.1 Principles

Smart antenna technology covers the smart antenna array design and smart antenna selection algorithm used to select antennas from antenna arrays.

Beam switching is essential for an antenna array. It ensures that multiple antennas are smartly selected from the antenna arrays of multiple antenna hardware units for signal sending and reception. This way, signal coverage in different directions is provided with different antenna combinations to deliver optimal signal quality to STAs at different locations, improving overall throughput. Beam switching is illustrated in Figure 3.75.

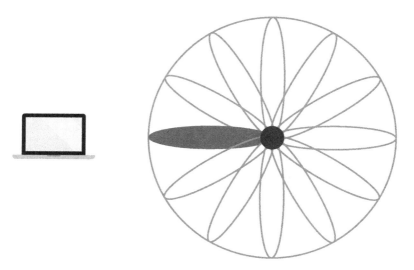

FIGURE 3.75   Beam switching.

The smart antenna selection algorithm is essential for link adaptation in WLANs. Based on STA locations, the algorithm selects the most appropriate combination from antenna arrays for data packet transmission, thereby improving overall WLAN performance. With smart antennas, directional beams are used to replace omnidirectional beams to concentrate beam energy toward target STAs, improving signal quality and overall system throughput.

Huawei's smart antennas adopt dual-band dual-polarization design. Dual-polarization design ensures miniaturized size, saving space and simplifying AP design. Dual-band design reduces inter-antenna coupling to increase radiation efficiency. As illustrated in Figure 3.76, four antennas, including two horizontally polarized antennas and two vertically polarized antennas, are provided on each of the 2.4 and 5 GHz bands.

Four beamforming structures are evenly distributed within the 360° range on the horizontal plane of each antenna. This ensures that the switch of each beamforming structure can be independently controlled. By turning on the switches in different directions, directional beams are obtained in different directions. By contrast, when all the switches are turned off, an omnidirectional beam is provided. Therefore, each antenna supports a total of five beamforming modes, as illustrated in Figure 3.77. With four antennas in place, a total of 625 combinations are supported.

- One mode (No. 2) for omnidirectional beams

- Four modes (No. 0, No. 1, No. 3, and No. 4) for directional beams in four horizontal directions

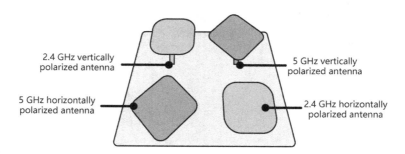

FIGURE 3.76   Layout of a smart antenna array.

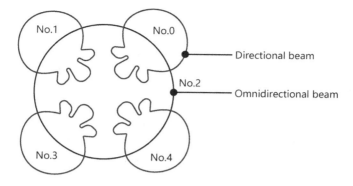

FIGURE 3.77   Five beamforming modes supported by a smart antenna.

3.7.2.2 Beam Selection

The smart antenna selection algorithm is essential for smart antennas. Based on the algorithm, training packets are sent using current antenna configuration, and the most appropriate antenna configuration is selected for the user based on the feedback of the PER and RSSI at the bottom layer obtained using the current antenna configuration. An antenna configuration mainly includes the combination of antennas and transmission rates.

As illustrated in Figure 3.78, an antenna has three working states: default, pretraining, and training. The smart antenna selection algorithm is implemented based on transition between these three states.

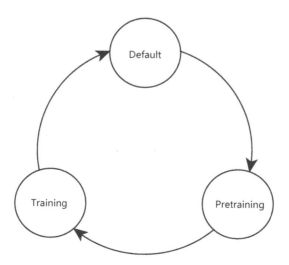

FIGURE 3.78   Transition between antenna working states.

In the default state, the antenna operates normally based on the current configuration and does not perform training operations. The system detects the antenna's working state and triggers training operations upon receiving an antenna training triggering request or sensing a sudden performance change.

In the pretraining state, the training configuration is provided through implementation of the rate adaptation algorithm. The training request is set and training is triggered only when the number of training packets is no smaller than that specified by the system. Antennas enter this state to reduce training time and channel occupation.

In the training state, the system traverses all antenna combinations based on the antenna configuration feedback to find the optimal antenna combination.

3.7.2.3 Applications of Smart Antennas

Smart antennas are widely used to improve WLAN performance. They are applicable to medium- and long-distance coverage scenarios, complex radio environments, high-density scenarios with multiple concurrent users, and scenarios with strong signal interference.

1. Medium- and long-distance coverage scenarios

    When an AP sends data to STAs, the smart antenna algorithm selects the most appropriate directional beam to replace the omnidirectional beam based on the locations of STAs. Directional beams allow for high gains, improving the coverage for STAs that are located at a medium or long distance from the AP, as illustrated in Figure 3.79.

2. Complex radio environments

    The antenna array of smart antennas provides better coverage in specified target areas. For example, in areas where signals penetrate floors or walls, the high gains of directional beams provide smart antennas with obvious penetration advantages. In the case of penetration failure, smart antennas select other directional beams to bypass the obstacles through reflection and diffraction. Figure 3.80 illustrates the effect of smart antennas in complex radio environments.

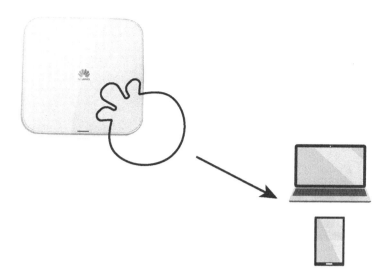

FIGURE 3.79   Beam selection for medium- and long-distance coverage.

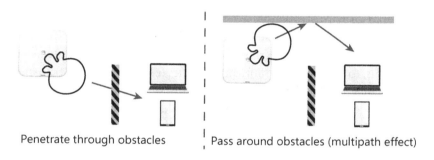

Penetrate through obstacles   Pass around obstacles (multipath effect)

FIGURE 3.80 Beam selection in environments with obstacles to signal propagation.

3. High-density scenarios with multiple concurrent users

In high-density scenarios, concurrent downlink transmission (including downlink MU-MIMO) is common. The directional beam selection of smart antennas allows signals of STAs in the same direction to aggregate and use the same directional beam for transmission. In doing so, the RSSI is improved and the interference between STAs in different directions is reduced. Figure 3.81 illustrates the effect of smart antennas in high-density scenarios with multiple concurrent users.

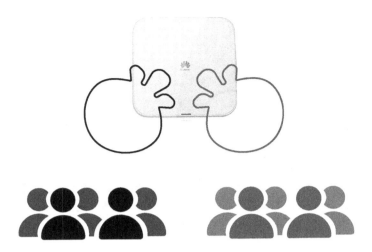

FIGURE 3.81 Downlink transmission in high-density scenarios with multiple concurrent users.

4. Scenarios with strong interference

The smart antenna selection algorithm selects beams that have high gains to desired signals and low or negative gains to interfering signals to enhance the anti-interference effect of uplink transmission. Beams that have maximum gains for receiving STAs can be selected in downlink transmission so that the high gains of directional beams are fully leveraged to mitigate interference and improve the anti-interference effect. Figure 3.82 illustrates the effect of smart antennas in scenarios with strong interference.

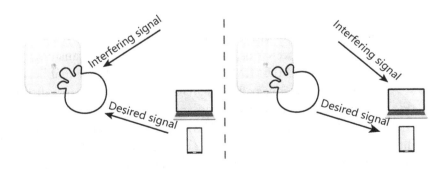

FIGURE 3.82 Anti-interference effect of smart antennas

3.7.3 High-Density Antenna Technologies

In high-density scenarios, a large number of users gather in a specific area, such as a conference room, classroom, lecture hall, stadium, and conference hall. In these scenarios, high-density antennas are an advanced and practical antenna technology used to achieve comprehensive Wi-Fi coverage.

Such high-density scenarios feature a high number of users per unit area. This requires small-angle directional antennas to be used in Wi-Fi networks to reduce the coverage radius of each AP for ensured overall capacity. As illustrated in Figure 3.83, with limited channels and small spacing between co-channel APs, the interference between co-channel APs substantially decreases the SNR within the coverage area. As a result, the communication quality deteriorates. Therefore, suppressing the side lobes of directional antennas is essential to suppressing the co-channel interference on high-density Wi-Fi networks.

In high-density Wi-Fi networks where APs are installed high above the ground (e.g., 15–50 m above the ground in sports stadiums), installation is difficult and requires coordinated operations, significantly increasing deployment costs. As Wi-Fi networks traditionally use external antennas in most cases, installation requires complicated auxiliaries for AP antennas and cables, making even more inconvenient. In dual-band high-density Wi-Fi networks, in addition to the mounting kits of APs, two to three mounting kits are also needed for antennas, and antenna downtilt adjustment needs to be completed.

To address the abovementioned challenges, mature high-density antenna solutions are already available in the industry. For example,

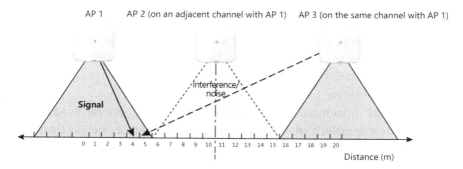

FIGURE 3.83   Co-channel interference in high-density Wi-Fi networks.

Huawei's all-in-one AP product features integration with high-density antennas. It incorporates a number of leading technologies, including Huawei-proprietary dual-band co-port planar array antennas, hyper-surfacing of artificial magnetic conductors, and parasitic radiation suppression in feeding networks. It allows for simplified deployment and ensures excellent side lobe suppression.

3.8 AI-POWERED IMPROVEMENT IN RADIO PERFORMANCE

As new radio access technologies and solutions continue to emerge, the differences in traffic created by existing and new services are extensive. These new services pose higher network performance requirements, resulting in larger and more complex networks. This also symbolizes the existence of an inevitable trend where intelligent technologies are being introduced, especially those based on artificial intelligence (AI), to develop future telecommunications networks.

Taking radio calibration as an example, the deployment as well as operation and maintenance phases of this process traditionally is labor-intensive and time-consuming. Generally, a time period of 3–5 weeks is required to complete the planning, deployment, commissioning, and acceptance procedures for a network that has thousands of APs in a typical large office campus. This is because traditional approaches are based on floor-by-floor commissioning, inevitably requiring manual intervention when APs are deployed on different floors.

Another disadvantage is that it takes a significant amount of time to resolve issues found during calibration. Furthermore, after a Wi-Fi network is deployed, the radio environment might change with time due to, for example, office reconstruction and more STAs accessing the network. As a result, network quality and user experience both deteriorate. Based on radio frequency (RF) detection for sensing radio environments, traditional approaches are too periodic and passive to actively respond to service changes. This is because only RF interference signals are considered during the calibration process.

Huawei has introduced big data technology to improve radio calibration. This new approach enables WLAN channel design as well as simulation feedback to be complete within seconds, and also provides optimal calibration suggestions, minimizing time spent and reducing possible errors. In addition, historical data can be considered to help locate edge

APs and predict AP load trends, enabling predictive calibration on wireless devices. To create a process that is more intelligent in its entirety, Huawei's big data-based intelligent approach optimizes data from the following aspects:

- Big data-based topology collection: Network topologies are collected through radio scanning after radio calibration starts. While radio calibration is being conducted, APs will randomly leave their working channels to scan other APs' working channels. Therefore, it is likely that two APs scan each other's working channels simultaneously, resulting in the two APs failing to discover each other as neighbors. Consequently, the network topology will be incomplete or inaccurate. Big-data based topology collection enables network topology to be gradually collected over a long period of time, ensuring the authenticity and reliability of topology data while also enabling network adjustment based on accurate network topology.

- Historical interference data collection: Frequent radio calibration impacts normal service traffic. Due to the weak compatibility and protocol compliance of some APs, it is likely that STAs will undergo service drops after channel switching on the APs. To minimize possible impact, routine radio calibration during operation and maintenance is usually performed with a long interval or a specified time. Specifically, for radio calibration performed at night, due to relatively light network traffic, interference sources can no longer be detected since the real traffic load created during the day is nonexistent. This compromises the radio calibration effect. By taking historical data into account, the real traffic load can be restored, providing accurate data associated with radio calibration.

- Network traffic prediction: Radio calibration is based on service traffic, and its main aim is to eliminate interference. This means that minimized interference specifically adapted to current traffic load does not necessarily guarantee future optimal service experience. Intelligent radio calibration includes a process of assessing historical data, such as service traffic and STA quantity. The new approach meets service requirements by combining radio calibration with real-life service traffic.

- Edge AP recognition and handling: STAs accessing the APs at the edge of a logical or physical topology are usually nomadic terminals. These STAs are unable to perform services and much likely to be disconnected. If there are too many nomadic STAs on the network, communication quality will be compromised. Therefore, effective recognition measures are required to locate edge APs and perform special handling upon them, limiting the impact on the network.

Wi-Fi will further integrate with AI technology to learn STA load distribution, roaming trajectory, access delay, and other information from a massive scale of historical running data. This is conducive to helping optimize air-interface performance while also making networks more intelligent.

CHAPTER 4

WLAN Security and Defense

W IRELESS LOCAL AREA NETWORKS (WLANs) are easier to build than wired networks as they transmit data over radio waves rather than through cables, but it is a challenging task to ensure security for wireless data transmission. This chapter begins with the security threats WLANs may encounter. We then proceed to common security mechanisms used for eliminating the threats, such as common access authentication modes as well as wireless attack detection and countermeasure.

4.1 WLAN SECURITY THREATS

The WLAN market demand keeps increasing, yet some potential users still think WLAN is not reliable enough. Compared to wired networks using cables (copper twisted pairs or optical fibers), WLANs transmit data over radio waves in the air, which is an open channel that can be exploited by attackers to intercept and even tamper with transmitted data.

Some of the most frequently encountered security risks include:

1. Occupation of authorized users' bandwidth by unauthorized users

    A simple password (such as the default password or one that is comprised of digits only) can be guessed or cracked by brute force so that attackers can access the network and occupy authorized users' bandwidth. Network service providers, therefore, advise home users to use Wi-Fi Protected Access 2 (WPA2) authentication and complex passwords (such as one that contains digits, letters, and special characters). Enterprise users can deploy the Secure Sockets Layer (SSL), which performs two-way certificate authentication to improve security.

2. Personal data theft by phishing websites

In cellular networks, attackers can exploit the intrinsic vulnerabilities of GSM technologies and, in disguise of rogue base stations, send junk data such as fraud information and push advertisements to users' mobile phones. Similarly in WLANs, particularly in public places, an unauthorized user can disguise as a dummy Wi-Fi hotspot with the same or similar service set identifier (SSID) as an authorized hotspot. When users access the dummy hotspot, important data may be intercepted, causing financial losses. For example, if a user performs a transaction on a shopping website advertised by the dummy hotspot, the user account can be hijacked.

Also, the intrinsic vulnerabilities of old standards, such as insecure algorithms and weak initialization vectors (IVs) used by WEP, make WLAN passwords more prone to cracking. Early in 2005, engineers used publicly available tools to crack a network using wired equivalent privacy (WEP) within 3 minutes. The Key Installation Attack (KRACK) widely reported in the second half of 2017 triggered a wave of doubt about wireless network security, particularly about Wi-Fi Protected Access (WPA) and WPA2.

In his work *Key Reinstallation Attacks: Forcing Nonce Reuse in WPA 2*, Belgian security researcher Mathy Vanhoef pointed out that the logical defects in WPA and WPA2 can trigger the reinstallation of keys, which provides attackers with an opportunity to decrypt wireless data packets through man-in-the-middle (MITM) attacks. Through exploiting this vulnerability, STAs can be forced to connect to unauthorized APs, which then extract credit card numbers, user names, passwords, photos, and other sensitive data from traffic streams. In some network configurations, attackers can even inject data into the network and remotely install malicious software, which will affect:

- STAs that use WPA2 authentication, such as mobile phones, tablets, and laptops;

- APs working as stations, for example, leaf APs in a wireless distribution system (WDS) or on a Mesh network.

Although the KRACK attack arises from a vulnerability in WPA_ Supplicant and increases the risk of attackers parsing the data packets sent in Linux systems and Android devices that use WPA_Supplicant, it does not crack the WLAN password. Still, Wi-Fi vulnerabilities caused users to worry about the existing wireless protocols, and so drove the Wi-Fi Alliance to come up with Wi-Fi Protected Access 3 (WPA3), the next-generation Wi-Fi encryption protocol.

In summary, common security threats to WLANs include:

1. Unauthorized use of network services

   Unauthorized users embezzle the WLAN bandwidth, which severely compromises authorized users' experience and may even cause serious damage in the event of malicious attacks.

2. Data security

   WLANs are more prone to user data interception and theft than wired LANs.

3. Rogue AP

   Rogue APs are an unauthorized existence in a WLAN and can obstruct normal network operations. They can be configured to capture STA data and enable access for unauthorized users, which can then capture data or send forged data packets and even access files on servers.

4. Denial of Service (DoS) attack

   DoS attacks are intended to stop the target device from providing services, and theoretically can be launched anytime during radio wave transmission. Such attacks are favorable to attackers because they are easy to launch and difficult to prevent.

To cope with such security threats, Huawei has designed security measures to protect user networks:

- Link authentication and user access authentication to prevent unauthorized use of network services: An enterprise-class user authentication solution (together with Huawei iMaster NCE-Campus) is deployed to centrally authenticate and manage user identities.

- Data encryption for data security: WPA3 with higher encryption strength is deployed to protect user data transmitted over the air interface from being cracked. WPA3 is one of the most powerful encryption algorithms, with a maximum key length of 256 bits.

- Wireless attack detection and countermeasure against rogue APs and DoS attacks: Huawei's Wireless Intrusion Detection System (WIDS) and Wireless Intrusion Prevention System (WIPS) are deployed to detect and counteract both air interface threats and rogue APs in real time, preventing unauthorized access to user networks.

In conclusion, effective security technologies can make enterprise WLANs just as secure as cellular networks. For details, Table 4.1 compares the security levels of enterprise WLANs and cellular networks.

4.2 WLAN SECURITY MECHANISMS

WLAN security mechanisms include link authentication, user access authentication, data encryption, and wireless attack detection and countermeasure.

1. Link Authentication

Link authentication is also referred to as STA identity authentication. The 802.11 standard necessitates link authentication before WLAN access. Therefore, link authentication begins when the handshake process starts for an STA to connect to an AP and then access a WLAN.

The 802.11 standard has defined two link authentication modes: open system authentication (OSA) and shared key authentication (SKA).

- In OSA, an STA is authenticated by its ID (usually a MAC address). All STAs that comply with the 802.11 standard can access a WLAN.

- SKA is supported only by WEP, requiring an STA and its serving AP to use one shared key.

WEP (and accordingly SKA) has been phased out due to poor security, and therefore OSA is used in link authentication. STAs can

TABLE 4.1   Security Comparison between Enterprise WLANs (Wi-Fi 6) and Cellular Networks (4G and 5G)

Security Factor		Wi-Fi 6 (WPA3)	5G	4G
Device identity		STA MAC address	IMSI[a] (globally unique identifier of a SIM card)	
Device identity confidentiality mechanism		MAC address randomization (using different random MAC addresses to access different SSIDs)	TMSI[a] (temporary IMSI, used for connection establishment)	
Terminal identification	Interworking NEs	STA and authentication server	UE and 3GPP authentication server	
	Identity	Enterprise-class certificate, user name/password	Identity authentication information in the SIM card	
Authentication	Authentication mode	EAP-TLS[a] (certificate), EAP-PEAP[a] (user name/password), and EAP-AKA[a] (using the SIM card)	3GPP and non-3GPP unified: 5G AKA and EAP-AKA'[a]	3GPP: EPS AKA[a] Non-3GPP: EAP-AKA and EAP-AKA'
Non-access stratum security algorithm	Encryption algorithm	AES	SNOW 3G, AES, ZuC	SNOW 3G, AES, ZuC
	Algorithm key	256-bit	256-bit	128-bit

[a] *Note*:
- *IMSI:* International Mobile Subscriber Identity
- *TMSI:* Temporary Mobile Subscriber Identity
- *EAP-TLS:* Extensible Authentication Protocol Transport Layer Security
- *EAP-PEAP:* Extensible Authentication Protocol-Protected Extensible Authentication Protocol
- *EPS:* Evolved Packet System
- *AKA:* Authentication and Key Agreement
- *EAP-AKA:* EAP Authentication and Key Agreement
- *EAP-AKA':* Improved EAP Authentication and Key Agreement

pass link authentication as long as protocol interaction is smooth, and therefore identity authentication does not really happen.

2. User Access Authentication and Data Encryption

In user access authentication, users are differentiated and their access rights are restricted before they eventually access a WLAN. User access authentication is more secure than simple STA identity authentication (or link authentication). Three modes for user access authentication are commonly used: WPA/WPA2+PSK authentication, 802.1X authentication, and WLAN Authentication and Privacy Infrastructure (WAPI) authentication (China National Standards).

Following on from user access authentication, data packet encryption is required to ensure data security. Encrypted data packets can only be decrypted by devices that have the key. Other devices are reluctant to do so, even if they receive data packets.

Currently, there are multiple data encryption methods, such as Rivest Cipher 4 (RC4), Temporal Key Integrity Protocol (TKIP), and Counter Mode with CBC-MAC Protocol (CCMP).

3. Wireless Attack Detection and Countermeasure

Authentication and encryption are commonly used wireless security solutions for protecting WLANs in different scenarios. In addition, wireless system protection technologies can also be leveraged, mainly WIDS and WIPS. The two technologies not only provide attack detection but also have countermeasures for more proactive protection.

4.2.1 Common WLAN Access Authentication Modes

4.2.1.1 WEP

WEP is an 802.11 protocol used to protect authorized users in a WLAN against data theft.

Its core encryption algorithm is RC4, a symmetric stream cryptographic algorithm that uses an identical static key for encryption and decryption. The encryption key can be 64, 128, or 152-bit long, with a system-generated 24-bit IV. In this way, the key length configured on a WLAN server and clients is 40, 104, or 128 bits.

WEP security mechanisms include link authentication and data encryption.

Link authentication is further categorized into OSA and SKA, as described in Table 4.2.

TABLE 4.2   Comparison between OSA and SKA

Authentication Mode	Key Negotiation Completed in Link Authentication	Service Data Encryption
OSA	No	Optional
SKA	Yes	Encryption with the negotiated key

4.2.1.2 WPA/WPA2

WEP SKA uses RC4 encryption which requires prior configuration of one static key, posing security threats to both the encryption mechanism and algorithm. To address this issue, the Wi-Fi Alliance launched WPA in 2003, as an enhancement to WEP.

WPA retains RC4 as the core encryption algorithm but introduces TKIP on the basis of WEP. WPA uses the 802.1X authentication framework and supports EAP-PEAP and EAP-TLS authentication. It uses a specific authentication server, usually a Remote Authentication Dial-In User Service (RADIUS) server, to implement two-way certificate authentication between the user and server.

In 2004, the 802.11i security standard organization launched WPA2, which uses CCMP encryption to deliver higher security than WPA.

4.2.1.3 WPA3

At the international Consumer Electronics Show (CES) held in Las Vegas on January 8, 2018, the Wi-Fi Alliance released WPA3, the next-generation Wi-Fi encryption protocol. On June 26, 2018, the Wi-Fi Alliance announced that WPA3 was finalized.

Table 4.3 compares technical details of WEP, WPA/WPA2, and WPA3.

TABLE 4.3 Technical Details of WEP, WPA/WPA2, and WPA3

Item	WEP	WPA/WPA2	WPA3
Application model	OSA: for telecom networks, using one-way STA authentication, network-layer portal authentication, or MAC address authentication SKA: for home networks, using two-way authentication, and requiring manual management of the initial key	Same as WEP, supporting OSA and SKA	OWE[a] can be used together with OSA link authentication to ensure data encryption. It is mainly used in wireless hotspots In the OSA scenario, OWE can be used together with OSA link authentication to ensure data encryption In the WPA3-Personal scenario, SAE[a] is forcibly used to replace WPA2-Personal authentication. SAE provides a transition mode that is compatible with WPA2-Personal In the WPA3-Enterprise scenario, SAE and OWE can be used to enhance security

(Continued)

TABLE 4.3 (*Continued*) Technical Details of WEP, WPA/WPA2, and WPA3

Item	WEP	WPA/WPA2	WPA3
Encryption	RC4 encryption, 40, 104, or 128-bit key, 24-bit IV, one key for both authentication and encryption	Enhanced encryption, 128-bit key, 48-bit IV TKIP: RC4 encryption, with increased key and IV lengths CCMP: AES encryption, with increased key and IV lengths, different keys (can be used unbound) for authentication and encryption	Only unified GCMP[a] encryption (i.e., AES-GCM[a]) supported, two password suite combinations GCMP-128, GMAC[a]-128, and SHA256[a] GCMP-256, GMAC-256, and SHA384
Integrity	CRC-32, simple data transmission error check mechanism, no integrity protection	TKIP: Michael algorithm to verify message integrity check information, integrity protection supported CCMP: CBC-MAC[a] mechanism, integrity protection supported, SHA1 hash algorithm	GMAC mechanism to implement integrity protection, SHA256 and SHA384 algorithms
Key negotiation and management	Static key, encryption using the master key	In both TKIP and CCMP, a dynamic master key can be generated, managed, and safely transmitted for each user through four-way handshake[a]. A temporary key derived from the master key, instead of the master key itself, is used for encryption	In GCMP, a dynamic master key can be generated, managed, and safely transmitted for each user through 4-way handshake. A temporary key derived from the master key, instead of the master key itself, is used for encryption

(Continued)

TABLE 4.3 (*Continued*)   Technical Details of WEP, WPA/WPA2, and WPA3

Item	WEP	WPA/WPA2	WPA3
Replay attack prevention	No protection mechanism	CCMP provides the packet number field to prevent replay attacks	GCMP provides the packet number field to prevent replay attacks
Management layer security	None	(Optional) PMF[a]	(Forcible) PMF
Privacy protection	None	None	(Optional) OWE, MAC address randomization, or SSID protection
Security suite B support (enterprise scenario)	None	EAP-PEAP and EAP-TLS	192-bit minimum security suite B (optional), with supporting GCMP-256, GMAC-256, and SHA384, and suite B TLS

[a] *Note*:
- *OWE:* Opportunistic Wireless Encryption
- *SAE:* Simultaneous Authentication of Equals
- *GCMP:* Galois-Counter Mode Protocol
- *AES-GCM:* AES with Galois/Counter Mode
- *GMAC:* Galois Message Authentication Code
- *CBC:* Cipher Block Chaining
- *PMF:* Protected Management Frame
- *SHA:* Secure Hash Algorithm
- *4-way handshake:* an EAPOL-Key negotiation mechanism (EAPOL is short for Extensible Authentication Protocol over LAN)

According to the Wi-Fi Alliance, WPA3-SAE will become a necessary function for Wi-Fi authentication after April 2020. Compared to WPA and WPA2, WPA3 is upgraded in the following respects:

- WPA3-SAE is introduced to provide a more secure handshake protocol. Theoretically, SAE can provide forward secrecy, making attackers reluctant to decrypt intercepted traffic even if the password is known. In a WPA2 network, however, traffic can be decrypted as long as the password is obtained. Therefore, WPA3-SAE has effectively addressed this issue.

- OWE is introduced to enhance user privacy in open networks by means of independent data encryption, allowing for unauthenticated encryption.

- Algorithms are improved. Security suite B is introduced, in which WPA3 supports AES-GCM with 256-bit keys or the 384-bit elliptic curve cryptographic algorithm.

4.2.1.4 WAPI

WAPI is a wireless security standard proposed by China based on the 802.11 standard, and is more secure than WEP and WPA. It consists of:

- WLAN authentication infrastructure (WAI): authenticates user identities and manages keys.

- WLAN privacy infrastructure (WPI): protects data transmitted on WLANs and provides data encryption, data authentication, and anti-replay protection.

WAPI uses the elliptic curve cryptographic algorithm based on the public key cryptography and the block cipher algorithm based on the symmetric cryptography. The two algorithms are used for the identification of digital certificates and other certificates and key negotiation between wireless devices, as well as the encryption and decryption of data transmitted between wireless devices. The two algorithms implement identity authentication, link authentication, access control, and user data encryption.

WAPI has the following advantages:

1. Two-way identity authentication

    It prevents access from unauthorized STAs to a WLAN and protects the WLAN against attacks from rogue devices.

2. Digital certificate as identity credential

    WAPI is backed by an independent certificate server. STAs and WLAN devices use digital certificates to prove their identities, improving network security. When an STA requests to join in or leave a network, the administrator only needs to issue a certificate or revoke the STA's certificate.

3. Robust authentication protocol

    WAPI uses digital certificates as user identity credentials, the elliptic curve cryptographic algorithm for authentication, and secure message authentication and hash algorithms to ensure data integrity. Attackers will find it difficult to modify or forge authenticated data due to high-level security.

4.2.2 WLAN User Authentication Modes

User authentication is an end-to-end network security measure to authenticate access terminals and users. WLANs generally use 802.1X authentication, MAC address authentication, portal authentication, WeChat authentication, and private pre-shared key (PPSK) authentication.

1. 802.1X authentication

802.1X authentication defined by the Institute of Electrical and Electronics Engineers (IEEE) is mainly used for Ethernet authentication and security.

The authentication system uses the client/server (C/S) structure and consists of three entities: applicant, authenticator, and authentication server, as illustrated in Figure 4.1.

In a WLAN, the applicant is a client (such as a mobile phone or a laptop) which must be able to support EAPOL on the WLAN. The authenticator is a wireless access controller (WAC) or an AP. It sends the authentication credentials submitted by the applicant to the authentication server and controls applicant access according to the authentication server's instructions. The authentication server,

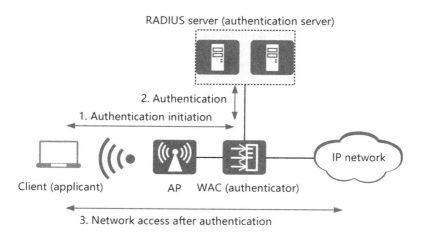

FIGURE 4.1   802.1X authentication system.

usually a RADIUS server, performs authentication, authorization, and accounting (AAA) on the applicant. The structure is more reliable because authentication really happens between the applicant and the authentication server, and the AP and WAC do not store the applicant's credentials. This facilitates unified management of authentication credentials on large-scale networks.

802.1X authentication uses the EAP protocol to implement data exchange across the applicant, authenticator, and authentication server. Common 802.1X authentication protocols include PEAP and TLS. Their differences are as follows:

- PEAP: The administrator assigns a user name and password which the user enters for WLAN access.

- TLS: Users are authenticated by their certificates. TLS authentication is normally used together with enterprise applications, such as Huawei AnyOffice.

802.1X authentication is recommended for large- and medium-sized enterprises.

2. Portal authentication

Portal authentication, also referred to as web authentication, is where a web browser suffices as the authentication client, as illustrated in Figure 4.2. Users must be successfully authenticated on the web portal before being allocated network resources. Service providers can leverage the web portal for services such as advertising.

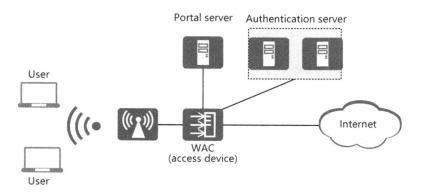

FIGURE 4.2   Portal authentication system.

Common portal authentication modes include:

- User name and password authentication: The foreground administrator applies for a temporary account to authenticate a guest.

- Short message service (SMS) authentication: Guests are authenticated by verification codes.

Portal authentication is recommended for guests of large- and medium-sized enterprises, business exhibitions, and public places.

3. MAC address authentication

It is usually used for dumb STAs (such as printers) to access a WLAN, or with the help of an authentication server to complete MAC address-prioritized portal authentication. After passing initial authentication, users can access a network again within a set period of time without being reauthenticated.

In this authentication mode, users' network access rights are controlled based on MAC addresses without needing to install a client, as illustrated in Figure 4.3. The access device starts authenticating a user immediately after detecting the user's MAC address for the first time on the interface where MAC address authentication has been enabled. During authentication, the user does not need to enter the user name or password.

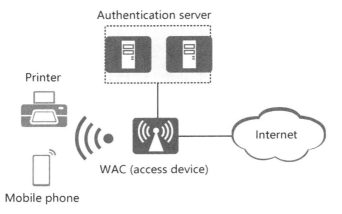

FIGURE 4.3   MAC address authentication system.

4. WeChat authentication

It is another common authentication mode, as illustrated in Figure 4.4.

WeChat is a free and popular application in China that provides instant messaging services on smart terminals. Its Official Accounts Platform provides merchants with a push advertising channel to do business. WeChat authentication is a special branch of Portal authentication, where users can get network services simply by following merchants' official accounts in an open network, without needing to enter their user name and password. Once the network is connected, users can browse the merchants' accounts and enjoy free Internet access.

5. PPSK authentication

It is a relatively secure authentication mode provided to employees in small- and medium-sized enterprises.

PPSK authentication is easier to deploy than 802.1X authentication, for it does not require a RADIUS server.

Compared with PSK authentication, PPSK authentication provides different pre-shared keys for different users, improving network security, as illustrated in Figure 4.5.

4.2.3 Wireless Attack Detection and Countermeasure

4.2.3.1 Basic Concepts

With wireless attack detection, WIDS detects attacks from unauthorized STAs and malicious users, and intrusions to WLANs. With wireless attack countermeasure, WIPS prevents unauthorized access to enterprise

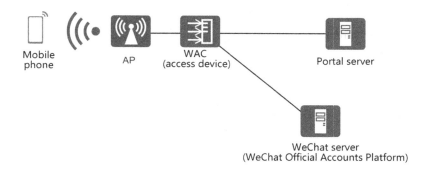

FIGURE 4.4   WeChat authentication system.

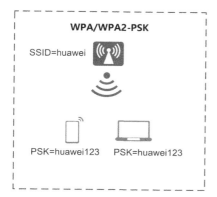
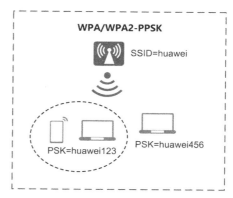

FIGURE 4.5   Comparison between PSK and PPSK authentication modes.

WLANs or authorized users from accessing unauthorized wireless devices. It further guards networks against attacks based on WIDS. The following are concepts related to WIDS and WIPS.

- Authorized AP: an authorized AP. For example, an AP managed by a WAC is an authorized AP for the WAC.

- Rogue AP: an unauthorized or malicious AP, which can be an AP that is connected to a WLAN without permission, an unconfigured AP, or an AP manipulated by an attacker.

- Interfering AP: neither an authorized AP nor a rogue AP. For example, when working channels of neighboring APs overlap, a highly powered AP causes signal interference and becomes an interfering AP.

- Monitoring AP: It scans or listens to radio media for attacks.

- Authorized wireless bridge: an authorized wireless bridge.

- Rogue wireless bridge: an unauthorized or malicious wireless bridge.

- Interfering wireless bridge: neither an authorized nor rogue wireless bridge.

- Authorized client: an STA connected to an authorized AP.

- Rogue client: an STA connected to a rogue AP. (Rogue APs, clients, and wireless bridges are collectively referred to as rogue devices.)

- Interfering client: an STA connected to an interfering AP.

- Ad-hoc STA: an STA that works in ad-hoc mode. Ad-hoc STAs can directly communicate with each other without the support of any device.

Based on the network scale, different features can be enabled for WIDS and WIPS:

- On home or small enterprise networks: blacklist and whitelist for access control of APs and STAs;

- On small- and medium-sized enterprise networks: WIDS for attack detection;

- On medium- and large-sized enterprise networks: detection, identification, and containment of rogue devices.

4.2.3.2 WIDS

WIDS can defend WLANs against attacks, such as 802.11 packet flood, spoofing attack, and weak IVs, and can shield brute force key cracking. When WIDS detects irregular behavior or packets, it determines that the network has been attacked and initiates security protection.

As illustrated in Figure 4.6, WIDS can be enabled when a WLAN is being accessed.

1. Flood attack detection

   During a flood attack, an AP receives massive amounts of the same type of management packets with the same source MAC address in just a short period. Such an attack prevents the AP from processing packets from authorized STAs due to its system resources being occupied with attack packets, as illustrated in Figure 4.7.

   Flood attack detection allows an AP to constantly monitor the traffic of each STA. When the traffic received from an STA exceeds the allowed threshold (e.g., over 100 packets per second), the AP considers that the STA will flood packets and reports an alarm to the WAC. If the dynamic blacklist function is enabled, the AP adds the detected attack STA to the dynamic blacklist. The AP then discards all the packets from this STA until the dynamic blacklist entry ages to prevent the network from a flood attack.

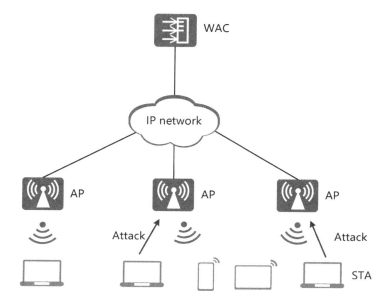

FIGURE 4.6   WIDS attack detection scenario.

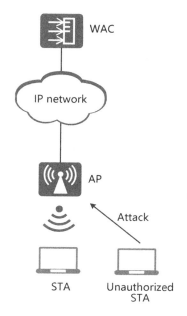

FIGURE 4.7   Flood attack principle.

An AP can detect flood attacks of the following frames:

- Authentication Request

- Deauthentication

- Association Request

- Disassociation

- Reassociation Request

- Probe Request

- Action

- EAPOL Start

- EAPOL-Logoff

2. Spoofing attack detection

In a spoofing attack (also called the MITM attack), an attacker (such as a malicious AP or user) masquerades as an authorized device and sends spoofing attack packets to an STA. As a result, the STA is denied access to the network, as illustrated in Figure 4.8. Spoofing attack packets include broadcast Disassociation and Deauthentication frames.

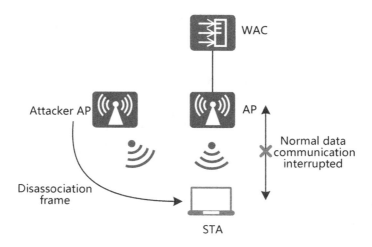

FIGURE 4.8   Spoofing attack principle.

Spoofing attack detection allows an authorized AP to, after it receives a Disassociation or Deauthentication frame, check whether the source IP address of the frame is the AP's MAC address. If so, the network is under a spoofing attack. In this case, the monitoring AP reports alarms to the WAC, which then notifies the administrator of the attack by means of recording logs or alarms.

Spoofing attacks cannot be prevented through the dynamic blacklist. This is because a rogue AP forges the MAC address of an authorized device, making it reluctant for the system to obtain the real MAC address of the rogue AP even if a spoofing attack has been detected.

3. Weak IV detection

As aforementioned, when WEP encryption is used on a WLAN, the system generates a 24-bit IV for each packet. In this case, when a WEP packet is sent, both the IV and the shared key are required to generate a key string which, together with the encrypted plaintext, forms the ciphertext. Weak IVs are generated using insecure methods, for example, repeated IVs are frequently generated or the same IV is always generated. When an STA sends a packet, the IV is sent in plaintext as a part of the packet header, which is a vulnerability that attackers can exploit to crack the shared key and then access the network, as illustrated in Figure 4.9.

Weak IV detection identifies the IV of each WEP packet to prevent such attacks. Then, when a packet is considered to contain a weak IV, the AP reports an alarm to the WAC so that users can use other security policies to prevent STAs from using the weak IV for encryption.

4. Defense against brute force key cracking

Brute force key cracking, an exhaustive attack method, is a cryptanalytic attack that tries every possible password combination to find the real password. For example, a password that contains only four digits may have a maximum of 10,000 combinations, and can therefore be cracked after a maximum of 10,000 attempts. In theory, it's just a matter of time that an attacker cracks any password, depending on the security mechanism and password length. In this sense, any authentication mode is vulnerable to brute force attacks.

When the WPA/WPA2+PSK, WAPI+PSK, or WEP+Share-Key security policy is used, brute force attacks may occur on the air

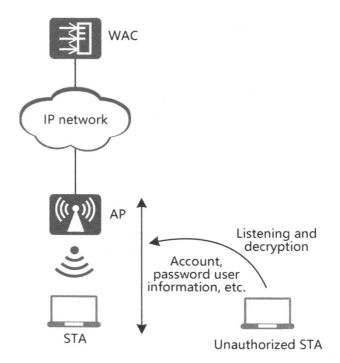

FIGURE 4.9 Weak IV attack principle.

Note: Weak IV detection does not require the dynamic blacklist function, as WEP authentication is rarely used nowadays due to high-security risks.

interface. In this case, to improve key security, PSK brute force cracking defense can be leveraged to prolong the time required for cracking a password. An AP checks the number of failed key negotiation attempts during WPA/WPA2+PSK, WAPI+PSK, or WEP+Share-Key authentication within a specified period. Then, if the number of failed key negotiation attempts exceeds a preset threshold, the AP suspects an STA is guilty of brute force cracking and sends an alarm to the WAC. If the dynamic blacklist function is enabled, the AP adds the STA to the dynamic blacklist and discards all packets from the STA until the dynamic blacklist entry ages.

PSK authentication and WEP+Share-Key authentication are deployed on the WAC and AP respectively, and therefore use different brute force attack detection methods. This is illustrated in Figure 4.10.

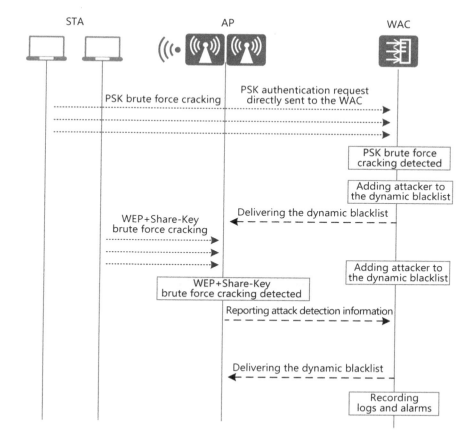

FIGURE 4.10   Example of detecting different brute force attacks.

4.2.3.3 Rogue Device Containment

In addition to defending against attacks on WLANs, WIDS also prevents unauthorized access to WLANs and STAs by rogue devices. Rogue device containment can identify and punish rogue devices to prevent them from gaining unauthorized access.

1. Rogue device identification

Upon receiving information about neighboring devices from an AP, the WAC starts rogue device identification. Figure 4.11 illustrates the process.

The WAC extracts neighbor information entries reported by APs one by one and performs the following judgment by device type:

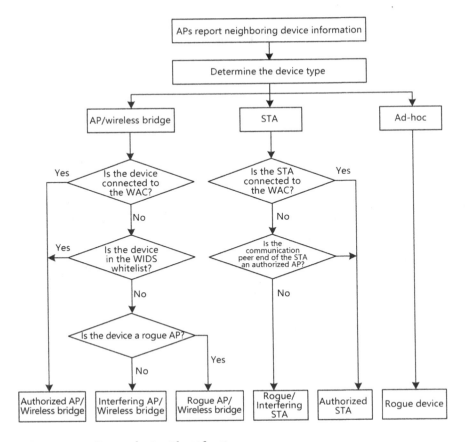

FIGURE 4.11   Rogue device identification process.

- If the device is an AP or a wireless bridge, the WAC checks whether the device is connected to the WAC. If so, the device is authorized. If not, the WAC checks whether the SSID, MAC address, or organizational unique identifier (OUI) of the device is in the WIDS whitelist configured by the administrator. If so, the device is authorized. If not, the WAC checks if the device SSID is the same as the local SSID or matches the spoofing SSID. If so, the WAC considers the device is a rogue device. In other cases, the WAC considers the device as an interfering device.

- If the device is an STA, the WAC checks whether the STA is connected to the WAC. If so, the STA is an authorized neighboring STA. If not, the WAC checks whether the communication peer end of the STA is an authorized AP. And, because the STA may be accessed using a whitelist SSID service, the WAC also checks the STA's basic service set identifier (BSSID). If the BSSID belongs to a member in the aforesaid authorized AP list, the STA is an authorized STA. In other cases, the STA is considered as a rogue or interfering STA.

- If the device type is ad-hoc, the WAC considers it as a rogue device.

2. Rogue device defense and containment

WLANs can defend against and counteract access from rogue devices, defense of which restricts access from rogue APs or STAs by configuring a blacklist. Rogue device containment allows an AP to download a contained device list from the WAC and take measures against rogue devices based on the device type.

To implement rogue device defense and containment (process illustrated in Figure 4.12), rogue device detection and identification need to be configured on an AP.

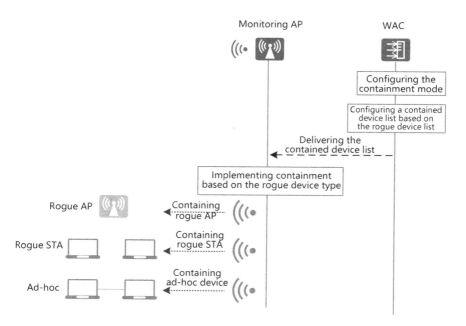

FIGURE 4.12   Basic process of rogue device containment.

In rogue device containment, the WAC configures and delivers a contained device list for the monitoring AP to implement countermeasures. The process is as follows:

- The WAC configures a contained device list based on the administrator-specified device type and enables the containment function.

- Based on the configured containment mode, the WAC selects a list of contained devices from the wireless devices reported by the monitoring AP each time and delivers the list to the monitoring AP.

- The monitoring AP takes countermeasures against the contained devices.

The countermeasures are as follows.

- Rogue or interfering AP: Once determining that an AP is a rogue or interfering AP, the WAC notifies the monitoring AP of the rogue or interfering AP. The monitoring AP broadcasts a Deauthentication frame as a rogue or interfering AP. Upon receiving the Deauthentication frame, STAs that have accessed the rogue or interfering AP disconnect from the rogue or interfering AP. In this way, STAs are prevented from associating with rogue or interfering APs.

- Rogue or interfering STA: Once determining that an STA is a rogue or interfering STA, the WAC notifies the monitoring AP of the STA. Then, the monitoring AP unicasts a Deauthentication frame as a rogue or interfering STA. Upon receiving the Deauthentication frame, the AP that is accessed by the rogue or interfering STA disconnects from the rogue or interfering STA. This way, the mechanism can terminate the connection between an AP and rogue or interfering STAs.

- Ad-hoc device: Once identifying an ad-hoc device, the WAC notifies the monitoring AP of the device. Then, the monitoring AP unicasts a Deauthentication frame as the ad-hoc device (by using its BSSID or MAC address). Upon receiving the Deauthentication frame, STAs that have accessed the ad-hoc device disconnect from the ad-hoc device. This way, the mechanism prevents STAs from associating with ad-hoc devices.

CHAPTER 5

WLAN and IoT Convergence

THIS CHAPTER BEGINS WITH wireless communications technologies commonly associated with Internet of Things (IoT). We then proceed to the feasibility of converged deployment of short-range wireless communications technologies used by Wi-Fi and IoT. Finally, several typical application scenarios of IoT APs are elaborated upon.

5.1 WIRELESS IoT TECHNOLOGIES

5.1.1 Overview of Wireless IoT Technologies

Almost in parallel to the rapid popularization of Wi-Fi, IoT is also experiencing fast development and wide application. IoT uses a broad array of wireless communications technologies, which roughly fall into short-range and long-range communication technologies, differing in their coverage capabilities. Common short-range wireless communications technologies include Wi-Fi, radio frequency identification (RFID), Bluetooth, and ZigBee. Common long-range wireless communications technologies include SigFox, Long-Range Radio (LoRa), and Narrowband IoT (NB-IoT).

1. RFID

    RFID is a wireless IoT access technology that can automatically identify target objects using radio signals and store and manage their information.

As illustrated in Figure 5.1, RFID works in four segments of frequency bands, namely low frequency (LF), high frequency (HF), ultra high frequency (UHF), and super high frequency (SHF). LF ranges from 120 to 134 kHz and supports a transmission distance of less than 10 cm. The HF band is 13.56 MHz and supports a transmission distance of less than 1 m. The UHF bands include 433, 865–868, 902–928 MHz, and 2.45 GHz, and support a transmission distance of 3–100 m. The SHF band is 5.8 GHz and supports a transmission distance of 1–2 m.

An RFID system consists of three parts: electronic tag, reader/writer, and information processing platform, as illustrated in Figure 5.2. The electronic tag communicates with the reader/writer through inductance coupling or electromagnetic wave reflection. Then, the reader/writer reads the electronic tag information and sends the information to the information processing platform over the network. Finally, the information processing platform stores and manages the information.

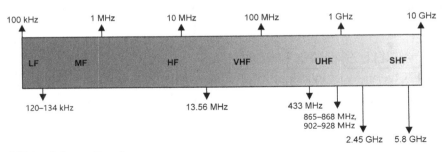

MF stands for medium frequency.
VHF stands for very high frequency.

FIGURE 5.1   RFID operating frequency bands.

FIGURE 5.2   RFID system.

RFID can be applied to a wide array of scenarios. Typically, LF RFID devices are used for access control and attendance card swiping, HF RFID devices are used for smart shelves and book management, and UHF RFID devices are used in supply chain management and logistics management scenarios.

2. Bluetooth

Bluetooth is one of the most widely used short-range wireless communications technologies. In 1994, Ericsson launched Bluetooth for the first time in the world. Then in 1998, Ericsson established the Bluetooth Special Interest Group (Bluetooth SIG) with Nokia, Intel, IBM, and Toshiba. Bluetooth SIG released the Bluetooth low energy (BLE) technology to satisfy fields such as smart home and fitness. Lately in 2016, Bluetooth SIG launched BLE 5.0, the next-generation Bluetooth standard which features faster transmission speed, longer coverage distance, and less power consumption.

BLE works on frequency bands from 2.4 to 2.4835 GHz, as illustrated in Figure 5.3, and occupies 40 channels, including three fixed broadcast channels and 37 frequency hopping data channels, with each channel occupying a bandwidth of 2 MHz.

With the emergence of IoT devices, including smart wearables, smart home, and Internet of Vehicles (IoV) in the IoT industry, Bluetooth is being increasingly used by developers. As a result, Bluetooth products including traditional Bluetooth-enabled mobile phones, headsets, speakers, mouses, and keyboards as well as new offerings such as smart bands, smart watches, vehicle-mounted devices, and smart home appliances are becoming more commonplace in the market.

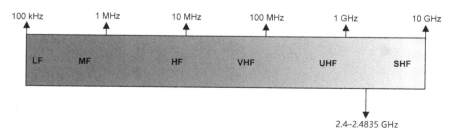

FIGURE 5.3   BLE operating frequency bands.

3. ZigBee

ZigBee is a self-organized short-range wireless communications technology that features low-power consumption and transmission rate. The ZigBee standard is released and maintained by the ZigBee Alliance, which industrializes stable, economical, low-power, and wireless surveillance products.

As illustrated in Figure 5.4, the operating frequency bands of ZigBee are distributed in 868, 915, and 2400 MHz, which respectively correspond to a channel bandwidth of 0.6, 2, and 5 MHz and a channel count of 1, 10, and 16. ZigBee transmits data at low rates, only 20 kbps in 868 MHz, 40 kbps in 915 MHz, and 250 kbps in 2.4 GHz.

As illustrated in Figure 5.5, there are three roles in ZigBee networking: a ZigBee Coordinator (ZC), a ZigBee Router (ZR), and a ZigBee End Device (ZED).

The ZC is the network coordinator responsible for establishing and managing the entire network. What's more, after the network is

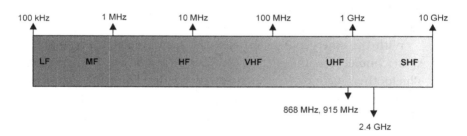

FIGURE 5.4   ZigBee operating frequency bands.

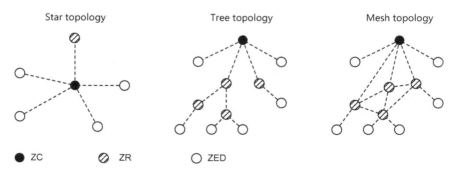

FIGURE 5.5   ZigBee networking.

set up, the ZC can also function as the ZR. The ZR provides routing information and allows other devices to join the network. The ZED acts as a client node that collects data.

ZigBee's main fields of application include: monitoring, sensing, automation, and control products in industry and agriculture; personal computers (PCs), personal digital assistants (PDAs), toys, and game controllers in consumer electronics; smart home such as security, lighting, and access control; personal care such as monitoring and medical diagnosis products.

4. SigFox

A wireless communications technology developed by SigFox, a French IoT company, SigFox uses Ultra Narrow Band (UNB) technology to serve low-rate machine-to-machine (M2M) applications.

As illustrated in Figure 5.6, SigFox works on 868 and 915 MHz, and provides a coverage distance of 30–50 km. It features low transmission rates and low-power consumption.

SigFox is mainly used in fields such as smart agriculture and environment, smart industry system, public utilities, smart home and life, smart energy system, and smart retail.

5. LoRa

LoRa is a wireless communications technology developed by Semtech that delivers long-distance communications and resists interference.

As illustrated in Figure 5.7, LoRa works on 433, 868, and 915 MHz, covers a distance of nearly 10 km, and provides a maximum transmission rate of about 50 kbps.

LoRa is mainly used in fields such as smart agriculture and environment, smart industry system, and public utilities.

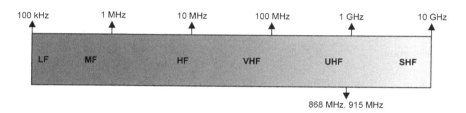

FIGURE 5.6   SigFox operating frequency bands.

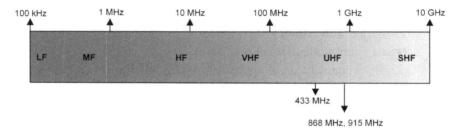

FIGURE 5.7   LoRa operating frequency bands.

6. NB-IoT

NB-IoT, whose definition and development are centrally governed by 3GPP, is targeted for low-frequency, small-packet, and delay-insensitive IoT services.

NB-IoT operates on the sub-1 GHz frequency band and features wide coverage, massive connectivity, low cost, and low-power consumption.

NB-IoT is mainly used for environment monitoring, agricultural production monitoring, smart city, and industrial equipment monitoring.

Tables 5.1 and 5.2 compare the foregoing wireless communications technologies.

TABLE 5.1   Comparison of Short-Range Wireless Communications Technologies

Item	RFID	BLE	ZigBee	Wi-Fi
Operating frequency band	120–134 kHz, 13.56, 433, 865–868, 902–928 MHz, 2.45 and 5.8 GHz	2.4–2.4835 GHz	868, 915 MHz, 2.4 GHz	2.4 or 5 GHz
Transmission rate	≤2 Mbps	≤25 Mbps	≤250 kbps	10 Gbps
Power consumption	≤1 W	≤100 mW	Sleep state: 1.5–3 μW Working: 100 mW	≤15 W
Coverage	10 cm to 100 m	10 m	10 m	15 m
Topology	Point-to-point, star	Point-to-point, star	Point-to-point, star, tree, and mesh	Point-to-point, star, tree, and mesh

TABLE 5.2   Comparison of Long-Range Wireless Communications Technologies

Item	SigFox	LoRa	NB-IoT
Operating frequency band	868 MHz 915 MHz	433 MHz 868 MHz 915 MHz	Sub-1 GHz
Transmission rate	0.1 kbps	0.3–50 kbps	0.44–195.8 kbps
Power consumption	50–100 μW	100 μW	100 μW
Coverage	Urban: 3–10 km Rural: 30–50 km	Urban: 1–1.8 km Rural: 4.5–9.2 km	Urban: 4.1–7.5 km Rural: 18.9–38.9 km
Topology	Star	Star or mesh	Star

5.1.2 Multinetwork Convergence of Wireless Communications Technologies

RFID, BLE, ZigBee, and Wi-Fi have similar coverage capabilities, as illustrated earlier in Tables 5.1 and 5.2. Convergence of these networks can significantly reduce the overall deployment and O&M costs. This section covers how to converge Wi-Fi networks with other short-range wireless communications technologies.

1. Multinetwork convergence trend

    In many projects, Wi-Fi covered areas require IoT access, too. For example, a large chain store needs to deploy both RFID and Wi-Fi networks for stock control and Internet access, and a large warehouse needs to deploy Wi-Fi and Bluetooth networks for Internet access and automatic-guided vehicle (AGV) positioning and navigation.

    Multinetwork convergence can reduce costs through co-site deployment as well as reduce maintenance costs through unified O&M, and is therefore helpful for controlling both capital expenditure (CAPEX) and operating expense (OPEX).

2. Deployment feasibility

    The coverage capabilities of wireless communications technologies vary. As a result, the distances between different wireless devices also vary. Co-site deployment is more feasible at shorter distances.

    As illustrated in Table 5.3, a Wi-Fi AP does not differ greatly from other types of base stations in terms of the coverage distance. The inter-site distance (ISD) of Wi-Fi APs is essentially the same as that of RFID base stations but is longer than that of Bluetooth or ZigBee base stations. In an ideal environment with low penetration loss

TABLE 5.3   ISD between Different Wireless Devices

Wi-Fi AP	Bluetooth Base Station	RFID Base Station	ZigBee Base Station
10–15 m	8 m	10–15 m	10 m

due to few spatial obstacles, co-site deployment of Wi-Fi APs is feasible. In other environments with high penetration loss due to severe blockage, Bluetooth or ZigBee base stations are required in addition to co-site deployment of Wi-Fi APs.

3. Feasibility of technologies coexisting on a single site

This section discusses the feasibility of adding other IoT access functions to an AP, thereby making an IoT AP.

Interference is a primary concern for the coexistence of multiple wireless communications technologies. Based on the characteristics of frequency bands, frequency band isolation can be used to prevent interference when a Wi-Fi AP coexists with an RFID or a ZigBee base station. For Bluetooth, frequency hopping can be used to avoid co-channel or adjacent-channel interference between the Wi-Fi and Bluetooth networks on the 2.4 GHz frequency band. The Wi-Fi network can coexist with these three IoT technologies, as illustrated in Figure 5.8.

Another prominent issue is how Wi-Fi and other wireless communications technologies coexist. Generally, IoT APs serve as carriers and integrate other wireless communications technologies. Such integration is implemented either through built-in integration or external expansion for mainstream IoT APs. In built-in integration, an IoT AP comes with other RF modules in addition to the Wi-Fi module. In external expansion, however, an IoT AP provides external interfaces, such as mini Peripheral Component Interconnect Express (PCIe) and Universal Serial Bus (USB) interfaces.

5.2 APPLICATION SCENARIOS OF IoT APs

Huawei IoT APs leverage built-in integration to converge IoT with wireless local area networks (WLANs). Built-in integration has the following advantages over external expansion:

- The IoT module is securely connected to an AP throughout the product lifecycle (5–8 years).

- The IoT module is embedded inside the AP, ensuring it does not compromise the AP's aesthetics or the ambient environments.

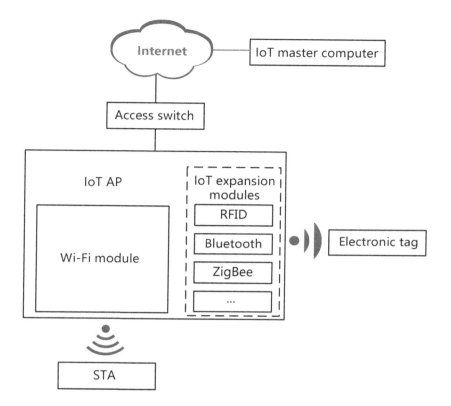

FIGURE 5.8 IoT AP.

This section uses three cases to illustrate the application scenarios of Huawei IoT APs.

5.2.1 Enterprise IoT

Enterprises need full-coverage Wi-Fi networks in office buildings to monitor important assets. Unique active BLE tags will be attached to important assets to identify them.

Take the asset management solution in Figure 5.9 as an example. In addition to providing Internet access for STAs, the APs also function as BLE base stations to periodically collect information on assets in BLE tags and send the information to the upper-layer IoT information control center. The IoT information control center analyzes data, obtains information on the location of the asset and its tracks, and displays the information on the GUI.

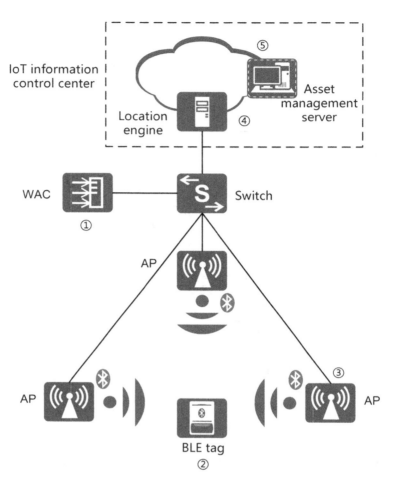

FIGURE 5.9   Enterprise IoT.

The operation process of the asset management solution is as follows:

1. The WAC manages APs and delivers asset management configurations to them.

2. BLE tags are associated with assets and periodically broadcast Beacon frames.

3. APs have built-in Bluetooth modules to listen to BLE tag broadcast frames and send information such as signal strength to the location engine.

4. The location engine processes the received information, analyzes assets' location information, and transfers the location information to the asset management server.

5. The asset management server displays the location of assets and provides asset management functions.

In this solution, the BLE tags used for the management of assets consume little power and have a battery life of 3–5 years. Management of assets is, in essence, an area positioning service that can determine which AP covers a BLE tag. Generally, positioning precision of an area is 20–30 m, which can meet enterprises' requirements for real-time monitoring of important assets.

When IoT APs are deployed and the Bluetooth function is enabled, channels 1, 6, and 11 are recommended for the 2.4 GHz frequency band. This is because BLE devices mostly transmit broadcast services, but the BLE protocol defines three 2 MHz broadcast channels that correspond to three fixed frequencies of 2402, 2426, and 2048 MHz. The three channels can be staggered with channels 1, 6, and 11 with center frequencies of 2412, 2437, and 2462 MHz, respectively, on the Wi-Fi network to avoid mutual interference.

5.2.2 Internet of Medical Things (IoMT)

Infant rooms in hospitals need an identity identification system for babies and a protection system for infants. This necessitates Wi-Fi in infant rooms to provide network access for medical staffs' PDAs. Figure 5.10 illustrates the IoMT solution that meets the requirements. In this solution, in addition to providing Wi-Fi access for PDAs, the APs also have RFID cards that connect to RFID security wristbands worn by the babies. After the baby wristband information is associated with the individual or primary caretaker's, such as the mother's, information on the healthcare management server, the healthcare management server obtains the locations of babies in real time through the APs. This way, the caretakers can confirm the identity and location of their babies on a mobile app. In addition, monitors as well as audible and visual alarm devices are deployed at the main exits and entrances in the hospital ward areas. Within the range of the baby wristbands, the monitors will trigger audible and visual alarms, ensuring infant security through a geofence.

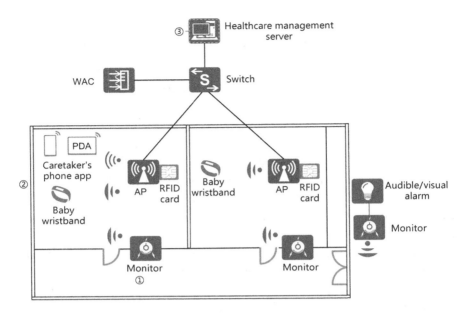

FIGURE 5.10   IoMT.

The operation process of the infant protection solution is as follows:

1. Monitors are deployed at the entrances and exits of key areas and periodically send Beacon frames.

2. Baby wristbands listen to the Beacon frames and send the monitoring information to the healthcare management server.

3. The healthcare management server matches the received information with the caretaker and infant information, and pushes the information to the mobile app of the caretaker.

5.2.3 Shopping Mall and Supermarket IoT

When deploying Wi-Fi in supermarkets, electronic shelf labels (ESLs) are used to replace traditional paper ESLs. This saves labor costs, as the server automatically updates ESL information such as commodity prices through the wireless network. ESLs can also display other information such as discounts and product specifications. As illustrated in Figure 5.11, the ESL management system connects to the customer's Enterprise Resource Planning (ERP) system to synchronize information such as commodity codes and prices.

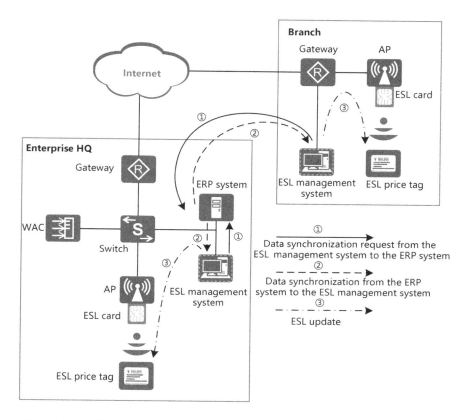

FIGURE 5.11 Shopping mall and supermarket IoT.

The operation process of the ESL solution is as follows:

1. The ESL management system server of each store requests data updates from the ERP system at the customer headquarters to obtain information on price changes.

2. After obtaining data, the ESL server delivers the ESL update task to the ESL card based on the task plan. The ESL card caches the data and waits for ESLs to initiate an update request.

3. An ESL is a low-power RFID terminal, with an RF module that wakes up periodically to query the ESL card for any ESL update tasks. If a task exists, the RF module obtains the data and refreshes the display; otherwise, the RF module enters the sleep state and waits for the next update task.

CHAPTER 6

WLAN Positioning Technologies

THIS CHAPTER BEGINS WITH the principles of common wireless positioning technologies. We then proceed to how various short-range communications systems analyze locations by using positioning technologies. Finally, this chapter introduces the concept of building a positioning system based on actual requirements.

6.1 DEVELOPMENT BACKGROUND OF WIRELESS POSITIONING TECHNOLOGIES

Developing positioning technologies has been a constant pursuit. From celestial navigation and lighthouses to satellite positioning systems, people have been benefiting profoundly from positioning technologies. For instance, when you first visit a city, you can quickly and accurately find your destination through the positioning service on your terminal's navigation app. Even in remote locations, accurate routes can be provided in just a touch of button on navigation apps.

In addition to conventional outdoor positioning services, people also have an urgent need for improvement in positioning services indoors in some scenarios, such as in large shopping malls, airports, office buildings, or hazardous production factories (where the real-time location of staff is needed). However, due to signal blockage by buildings and large objects, traditional outdoor positioning solutions are not suitable for indoor environments. As a result, various indoor positioning technologies have emerged that use short-range wireless communications technologies, such as Wi-Fi, Bluetooth, and Ultra-Wide Band (UWB), to provide positioning services.

6.2 TECHNICAL PRINCIPLES OF WIRELESS POSITIONING

Wireless positioning services locate the source and target based on the statistical characteristics of received radio signals. The main parameters of radio signals include the signal strength, angle of arrival (AoA), and propagation delay. As illustrated in Figure 6.1, wireless positioning falls into two categories: RF pattern matching and geometric positioning, which differ in their processing of data. The former includes RF pattern matching based on the received signal strength indicator (RSSI) and RF pattern matching based on the channel state information (CSI), while the latter includes trilateration, triangulation, and hyperbolic localization.

RF pattern matching associates a location with radio signal reception features at the location. This feature is unique to each location, similar to a fingerprint. The radio signal feature obtained in real time is matched with feature data in a preinstalled radio signal feature fingerprint database, and the location of a signal source is estimated based on the matching result. Conventional RF pattern matching solutions use RSSI as the signal feature to establish the fingerprint database, but there are cases in which CSI is used for this purpose.

Geometric positioning uses parameters such as AoA and propagation delay of radio signals to perform AoA-based triangulation, trilateration based on the round trip time (RTT) or RSSI, or hyperbolic localization based on the time difference of arrival (TDoA) to estimate the source of signals. These techniques were first used in radar systems. By processing

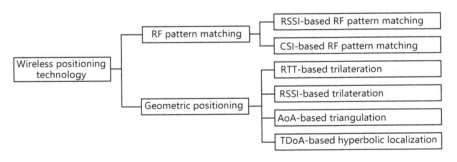

FIGURE 6.1 Categories of wireless positioning technologies.

the received signals in space, time, and frequency domains and with the help of effective positioning technologies, the target locations can be estimated and tracked.

6.2.1 RF Pattern Matching

RF pattern matching is divided into offline and online phases. As illustrated in Figure 6.2, in the offline phase (or training phase), a radio signal feature fingerprint database is established at the demarcated location. In the online phase, the fingerprint database is used to locate the target in real time.

After establishment, the fingerprint database needs to be updated in real time and maintained, usually through crowdsourcing. Crowdsourcing allows users to participate in the update and maintenance of the fingerprint database. They can modify the fingerprint database while enjoying the positioning service. This effectively keeps the fingerprints in the database updated to ensure accurate locations' information.

6.2.1.1 RSSI-Based RF Pattern Matching

RSSI-based RF pattern matching uses RSSI to establish the fingerprint database. As illustrated in Figure 6.3, a specified area is divided into several units by a specified granularity. In the offline phase, RSSI information is collected from some selected locations to establish a fingerprint database. In the positioning phase, the fingerprint database information is matched with the RSSI information collected in the online phase, with the closest location information collected offline in the fingerprint database selected as the output.

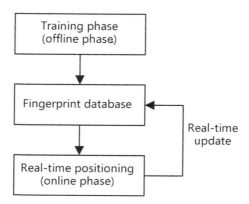

FIGURE 6.2   RF pattern matching.

(x_{00},y_{00})	(x_{01},y_{01})	(x_{02},y_{02})	(x_{03},y_{03})	(x_{04},y_{04})	(x_{05},y_{05})
AP 1	(x_{11},y_{11})	$(\cdots)$	AP 2	$(\cdots)$	$(\cdots)$
$(\cdots)$	$(\cdots)$	(x_{22},y_{22})	$(\cdots)$	$(\cdots)$	$(\cdots)$
$(\cdots)$	$(\cdots)$ AP 3 $(\cdots)$		(x_{33},y_{33})	$(\cdots)$	AP 4

Location Coordinates	AP MAC	RSSI
(x_{00},y_{00})	AP 1 MAC	−45 dBm
	AP 2 MAC	−65 dBm
	AP 3 MAC	−59 dBm
	AP 4 MAC	−70 dBm
(x_{01},y_{01})	...	...
...	...	...

FIGURE 6.3   Location collection in the offline phase.

The positioning precision of RSSI-based RF pattern matching is affected by multiple factors such as signal fluctuation, terminal location, and multipath transmission, which can result in relatively large errors.

6.2.1.2 CSI-Based RF Pattern Matching
CSI indicates the frequency domain response of a channel: specifically a channel gain feature corresponding to each frequency in a fixed frequency bandwidth at a location. As illustrated in Figure 6.4, channel gains differ

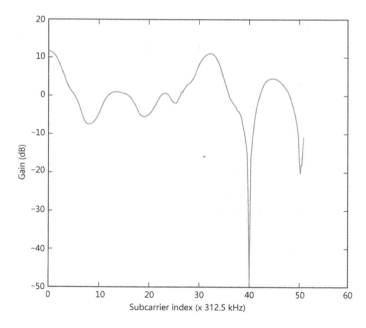

FIGURE 6.4   CSI information at a fixed point.

by frequencies. CSI-based RF pattern matching is used to establish a fingerprint database and match locations by using this channel gain feature. CSI-based RF pattern matching can accurately reflect small-scale features of a channel and provides significantly more detailed information than RSSI-based RF pattern matching. Therefore, the precision of CSI-based RF pattern matching is higher than that of RSSI-based RF pattern matching. A growing number of researches are focused on developing CSI-based RF pattern matching.

Similar to RSSI-based RF pattern matching, CSI-based RF pattern matching requires a fingerprint database to be established in advance and performs real-time positioning in the positioning phase using a classification algorithm. Even though that CSI-based RF pattern matching provides a large amount of channel information, a problem similar to RSSI-based RF pattern matching still exists in practice.

- CSI-based RF pattern matching obtains the small-scale information of a channel, but small-scale fading in a radio environment represents a channel feature of several wavelengths in a spatial range. To fully collect the channel's small-scale information, the granularity of a fingerprint must be within dozens of centimeters.

- Differences in the antenna directivity diagrams in terminals also affect a channel's small-scale fading. Due to differences in antenna directivity, the radiated energy of signals in different directions varies. As a result, the channel features change, affecting the positioning precision.

- Impact of the human body on a terminal also changes the CSI features.

6.2.2 Geometric Positioning

In a wireless communications system, because of the multipath effect, a received signal $y_m(t)$ is usually represented as a superposition of multiple signals after a transmit signal $s(t)$ arrives at a receiver through multipaths, as in the following formula:

$$y_m(t) = \sum_{i=1}^{L} \alpha_i e^{j2\pi f_{d,i} t} e^{j2\pi (\mathbf{d}_m)^T \alpha_i/\lambda} s(t - \tau_i)$$

The formula is defined as follows:

- L represents the quantity of multipaths;
- α_i represents the multiplexing gain of the ith path, and energy of α_i represents the path propagation gain;
- $f_{d,i}$ represents the Doppler frequency offset of the ith path;
- τ_i represents the propagation delay of the ith path;
- $\mathbf{a}_i$ represents the spatial feature vector of the ith path (specifically, it refers to a vector function a_i of the pitch AoA θ_i and the horizontal AoA φ_i, with $\mathbf{a}_l = g\left[\theta_i, \varphi_i\right]$);
- $\mathbf{d}_m$ represents the spatial location vector of the mth antenna;
- λ is the wavelength of an electromagnetic wave.

In geometric positioning, the received signal $y_m(t)$ is processed to obtain α_i, $f_{d,i}$, τ_i, θ_i, and φ_i so as to locate a signal source.

To better understand the technology, it is necessary to introduce the measurement method of the signal propagation distance and the AoA.

6.2.2.1 Propagation Distance Measurement

In a radar system, radio signals sent from the transmitter are reflected off the target and collected by the receiver. As illustrated by the system on the left in Figure 6.5, the propagation path length of radio signals is twice the distance between the target and the ground-based radar, assuming that the transmitter and receiver are deployed together. The system can calculate the propagation distance by measuring the signal RTT. Typical measurement methods include frequency-modulated continuous wave and pulse modulation.

In wireless local area network (WLAN) indoor positioning, because an AP and an STA exchange packets, the packet exchange time can be

FIGURE 6.5   Propagation distance measurement principle.

recorded to measure the propagation distance. In the system on the right in Figure 6.5, the AP sends a null frame to the STA, and the STA replies with an ACK frame. The AP records the time t_1 for sending the null frame and the time t_4 for receiving the ACK frame. It then calculates the packet RTT (Δt), with $\Delta t = t_4 - t_1 = 2 \cdot d$, where d represents the distance between the AP and the STA.

However, this can differ in a real environment, as depicted by the equation, $\Delta t = t_4 - t_1 = 2 \cdot d + \Delta t_{\text{delay}}$, where Δt_{delay} is the processing delay between the moment where the null frame arrives at the STA, and the moment where the STA replies with an ACK frame. Δt_{delay} fluctuates because the delay for each signal detection is different, affecting distance measurement precision. As a result, the distance is offset by 0.15 m for every 1 ns of fluctuation in delay.

To overcome this issue, the IEEE 802.11 standard group added the fine timing measurement (FTM) mechanism to the 802.11 revision released in 2016, to improve the distance measurement capability of Wi-Fi networks. The two ends of a link on which FTM measurement is performed serve as an initiator and a responder. They measure the distance between them by performing FTM-ACK interaction.

As illustrated in Figure 6.6, the responder records the sending time t_1 of the FTM frame and the receiving time t_4 of the ACK frame. The initiator records the receiving time t_2 of the FTM frame and the sending time t_3 of the ACK frame. In the next FTM-ACK interaction, the responder feeds back t_1 and t_4 to the initiator. In this manner, the initiator is able to calculate the processing delay $\Delta t_{\text{delay}} = t_3 - t_2$ by using t_2 and t_3.

It should also be noted that, because there are abundant multipath components in a radio channel in an indoor environment, fading caused by superimposition of multipath signals affects packet detection by the receiver. The positioning system focuses only on the propagation distance

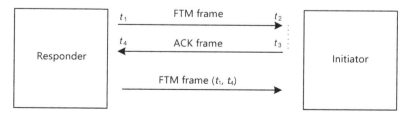

FIGURE 6.6   FTM principles.

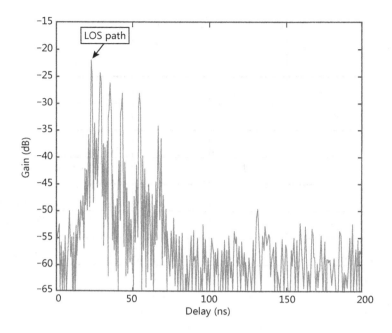

FIGURE 6.7   Relationship between multipath signals and delay.

of the signals from the line-of-sight (LOS) path, as illustrated in Figure 6.7. If signals from the reflection path are mistakenly considered as signals from the LOS path, the time of arrival (ToA) measurement error increases. To mitigate the impact of multipath signals on the ToA measurement, the multipath components are predominantly identified by means of difference or spectrum estimation. The propagation path from where signals first arrive is selected as the LOS path, and the ToA of the LOS path is obtained. The capability of differentiating multipaths depends on the radio system bandwidth. Therefore, the higher the bandwidth, the stronger the capability and the higher the time measurement precision.

6.2.2.2 AoA Measurement

The AoA measurement principle is similar to that in radar systems. To measure the AoA, the receiver must be equipped with an antenna array comprised of multiple receive antennas. The uniform linear array (ULA) is used as an example, as illustrated in Figure 6.8.

Assume that three receive antennas are evenly and linearly distributed in space at an interval of d. Signals from a remote terminal arrive at each

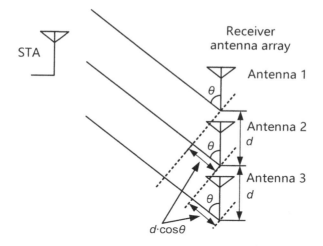

FIGURE 6.8   Antenna array.

receive antenna in the form of plane waves. The angle between the signal arrived and the linear array is θ. Assuming that the phases of the signals received by antenna 1 and 2 are $\varphi1$ and $\varphi2$, respectively, the phase of the signal received by antenna 2 $(2\pi d \cdot \cos\theta)/\lambda$ (λ is the wavelength of the electromagnetic wave) comes after the signal received by antenna 1. The phase difference depends on the signal AoA and the array structure. The receiver can measure the phase difference $\Delta\varphi$ between antennas. Therefore, according to $\Delta\varphi = (2\pi d \cdot \cos\theta)/\lambda$, when the antenna spacing d is known, the signal AoA θ can be obtained. Typical algorithms for calculating the signal AoA include Machine Utilization Statistical Information Collection (MUSIC) and Estimation of Signal Parameters via Rotational Invariance Technique (ESPRIT).

It should be noted that, similar to the propagation distance, the system only cares about the signal AoA from the LOS path.

The following typical geometric positioning technologies occur, after the measurement of signal propagation distance and AoA is introduced. The below describes these technologies.

6.2.2.3 Trilateration

Trilateration locates an STA by measuring the signal propagation distance. As illustrated in Figure 6.9, it measures the distances d_1, d_2, and d_3 between three APs and a target STA, to estimate the STA's location. Trilateration also falls into RTT-based trilateration and RSSI-based trilateration.

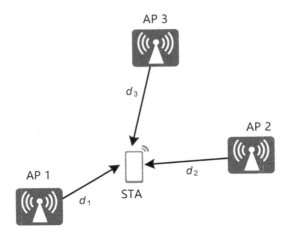

FIGURE 6.9   Trilateration principle.

1. RTT-based trilateration

   RTT-based trilateration measures the distance between an STA and an AP through packet exchange. As illustrated in Figure 6.10, the STA must exchange packets with the associated AP and the unassociated neighboring APs in sequence to obtain the distance to each AP. It then calculates the STA's location according to the known location of the AP by using the least square method. RTT-based trilateration does not require clock synchronization between APs.

2. RSSI-based trilateration

   RSSI-based trilateration uses the received signal strength and path loss model to calculate the distance between two ends. When shadow fading σ_{SF}^2 and small-scale fading σ_{ss}^2 are not considered, $RSSI = P_{Tx} - PL$. P_{Tx} represents the known transmit power, and PL is the path loss of signal propagation. In an ideal free space (FS), PL may be expressed as:

$$\text{FSPL}(\text{dB}) = 20\lg d + 20\lg f + 92.45$$

   where d is the distance between two ends of a link in a unit of km, and f is the center frequency of a signal in a unit of GHz. We can see that a stronger RSSI indicates a closer distance between two objects and a weaker RSSI indicates a longer distance. Therefore, the distance between two ends can be estimated by using RSSI. The path loss

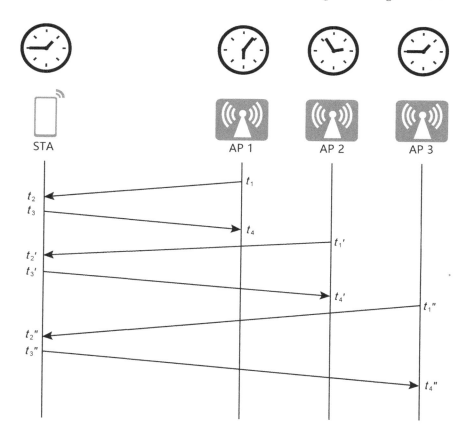

FIGURE 6.10   RTT-based trilateration procedure.

model varies depending on the environment, and therefore the path loss formula used is required to be modeled in the deployment phase.

6.2.2.4 Triangulation

The basic principle of triangulation is to measure the angle between the connection direction at two ends of a link and the reference direction. This is done to obtain the azimuth angle from an STA to at least three APs, and then estimate the STA's location, as illustrated in Figure 6.11.

In network-side triangulation, because only the AoA of an STA's uplink signals needs to be measured, an AP is only required to listen to uplink packets from the STA. This removes the requirement for packet inter-action. In addition, the AP must be configured with antenna arrays to measure the AoA of signals on the LOS paths at both ends of a link.

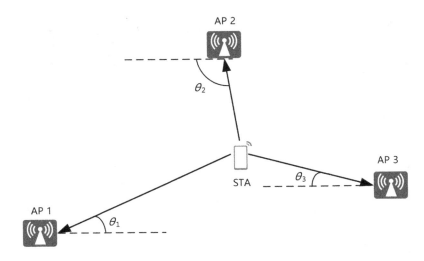

FIGURE 6.11   Triangulation principle.

6.2.2.5 Hyperbolic Localization

Hyperbolic localization calculates differences in signal propagation distances. This calculation is based on the time differences between uplink signals sent by a measured object to arrive at multiple measured objects. The STA location is then estimated based on the differences in signal propagation distances, as illustrated in Figure 6.12.

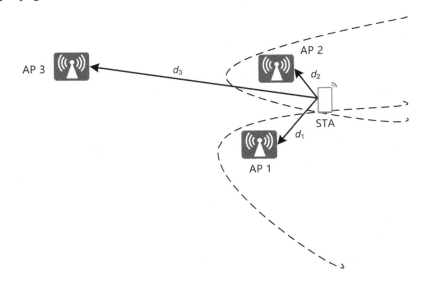

FIGURE 6.12   Schematic diagram of hyperbolic localization.

The propagation distance of uplink signals from an STA to each AP is different. If the distance difference between any two APs and the STA can be measured, for example, $\Delta d_{32} = d_3 - d_2$ and $\Delta d_{31} = d_3 - d_1$, the analytic expression of the STA location coordinates and the distance differences can be obtained. Then, the STA location can be calculated using the iterative Gauss-Newton algorithm. To calculate the distance difference, TDoA is predominantly used. As illustrated in Figure 6.13, the STA sends uplink packets to all neighboring APs. The APs synchronize their clocks with each other. The distance difference, for example, $\Delta d_{32} = d_3 - d_2 = c \cdot (t_3 - t_2)$, can be obtained by calculating the ToA difference between the APs.

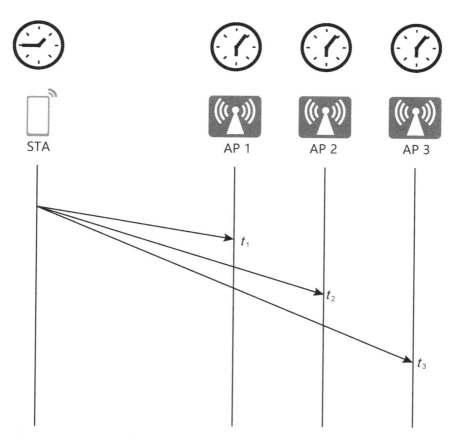

FIGURE 6.13   TDoA measurement process.

6.3 SHORT-RANGE WIRELESS POSITIONING TECHNOLOGY IMPLEMENTATION METHODS

Existing wireless communications technologies are typically used to estimate locations in the application of positioning technologies. This allows the management of location information for active devices in wireless positioning solutions. The following sections describe the applications of indoor wireless positioning technologies, in various short-range wireless communications systems.

6.3.1 Application of Indoor Positioning Technologies in Short-Range Wireless Communications Systems

The short-range wireless communications systems mentioned herein refer to systems that use radio transmission of different standards, including Wi-Fi, Bluetooth, and UWB.

6.3.1.1 Wi-Fi

In Wi-Fi systems, the most common positioning technology is RSSI-based positioning. Specifically, an AP measures STA locations. Upon receiving uplink signals from an STA, the AP estimates the RSSI and sends the RSSI to the location engine. The location engine then calculates the STA location based on the trilateration principle. Alternatively, the AP sends Wi-Fi signals to an STA and then the STA, upon receiving the signals, parses the RSSI and finally calculates its own location. The RSSI values measured in a Wi-Fi system fluctuate greatly, due to various factors such as complex spatial environments. This compromises the positioning precision which can only reach 3–7 m.

To improve the positioning precision, vendors use RF pattern matching as an auxiliary measure. After an AP has been deployed, its coverage area is divided into multiple location grids, and the coordinates of each grid are determined. Then, a test STA collects the RSSI features corresponding to each grid and sends them to the fingerprint database. This way, the fingerprint database is established. During positioning, upon collecting the RSSI of uplink signals from the test STA, the location engine matches the RSSI against those in the fingerprint database and selects the location grid coordinates corresponding to the best-matched RSSI feature as the STA's final location. This increases positioning precision to 3–5 m. However, the RF pattern matching is rarely used. For one thing, it is costly to deploy and the AP location cannot be randomly changed after deployment. In addition, this method is sensitive to STA differences and maintenance of the fingerprint database further increases costs.

As a new indoor positioning technology, AoA-based positioning is not subject to signal strength fluctuation and delivers significantly higher precision than RSSI-based positioning. AoA-based positioning does not require path loss model evaluation or a fingerprint database, nor is it subject to transmit power restrictions. This reduces network deployment and maintenance costs. This solution uses the in-phase/quadrature (I/Q) data or CSI collected by chip hardware and the array signal processing technology to achieve positioning precision of 1–3 m in indoor environments. However, because an array antenna of a specific structure is required to estimate the signal azimuth angle, hardware costs of this solution are much higher than those of RSSI-based positioning.

Given that the technology has been released by the 802.11 standard and is supported by mainstream chip vendors, FTM has been attracting industry-wide attention. Due to the fact that Wi-Fi works on 20, 40, 80, and 160 MHz bandwidths, FTM uses a multicarrier bandwidth system to implement high-precision measurement of the signal receiving time, thereby achieving high-precision distance measurement. At 80 MHz system bandwidth, FTM can deliver distance measurement precision of 1–2 m, making meter-level indoor positioning possible. With the popularization of FTM-capable terminals, FTM-based indoor positioning solutions will be used in a variety of applications.

6.3.1.2 Bluetooth

The mainstream Bluetooth positioning solutions are similar to Wi-Fi RSSI-based positioning solutions, namely RSSI-based RF pattern matching and trilateration. With the release of the BLE standard, low-power tag devices can be powered by a battery, with a battery life of several years. This makes Bluetooth positioning ideal for asset management. BLE-capable Bluetooth modules can be integrated into network devices to collect, through frequency hopping, on three broadcast channels, the uplink signals sent by BLE tags and measure the RSSI. Finally, the Bluetooth modules determine the BLE tag location based on an estimated distance. BLE-based positioning precision can reach 3–5 m, which is higher than that of Wi-Fi RSSI-based positioning.

In addition, the iBeacon function released by Apple in September 2013 can be used to locate Bluetooth-enabled smart terminals. Battery-powered anchors are fixedly distributed in an indoor environment, with a deployment density usually higher than that of Wi-Fi devices. Smart terminals listen to packets sent by surrounding anchors on the Bluetooth broadcast channel, and calculate and update their locations in real time based on collected RSSI information.

6.3.1.3 UWB

UWB systems predominantly use two positioning technologies, namely trilateration and hyperbolic localization. With the ultra-large bandwidth of 500 MHz or 1 GHz, time measurement precision can reach 100 ps, enabling distance measurement precision to be maintained at 10–20 cm. RTT-based trilateration requires that an STA separately exchanges packets with multiple anchors to obtain the distances between the STA and the anchors. This requires a large quantity of air interface resources and can only locate a limited number of STAs. Therefore, TDoA-based hyperbolic localization is often used in practice. Time between anchors is synchronized generally through wired or wireless synchronization. Wired synchronization uses an external clock source to provide time for all anchors on a network. This mode is costly and is not suitable for large-scale networks. Wireless synchronization implements clock synchronization between anchors through packet exchange over the air interface, simplifying the deployment solution.

6.3.2 Evaluation Counters of Indoor Positioning Solutions

6.3.2.1 Precision

Positioning precision refers to the degree of proximity between location information (usually coordinates) and the real location of a spatial entity. Precision of wireless positioning depends on three aspects:

1. Environment

   Unpredictability of the radio environment leads to positioning uncertainty. Multipath effect, blocking of human bodies and walls, and interference from other radio signals affect the measurement of parameters, such as RSSI, AoA, and ToA, compromising positioning precision. A single technology can barely achieve high-precision positioning across an area. In this sense, positioning solutions backed by multiple technologies are effective in mitigating the impact of environmental factors.

2. Bandwidth/Antenna quantity

The radio system bandwidth determines ToA measurement precision. For example, UWB can achieve centimeter-level precision by virtue of large bandwidths, while Wi-Fi can achieve only meter-level because devices usually work at a bandwidth of 40 or 80 MHz. In addition, the number of antennas in an antenna array determines AoA measurement precision. Similar to beamforming where the beams are finer if the array apertures are larger, AoA measurement precision is higher if there are more antennas.

3. Demarcation precision

A prerequisite for STA positioning is that the locations of network devices are known, which means the network devices must first be demarcated during network deployment. Demarcation precision indirectly affects positioning precision. In trilateration and hyperbolic localization, accurate AP coordinates need to be recorded. For triangulation, accurate AP coordinates and antenna array direction need to be recorded.

6.3.2.2 Power Consumption

Power consumption reflects the endurance of terminals to be located and is mainly determined by the modulation scheme and the refresh rate of radio signals.

1. Modulation scheme

In wireless communications systems, the power amplifier (PA) of a transmitter consumes the most energy, and PA efficiency depends on the radio signal modulation scheme. For constant envelope modulation, the PA is in a linear area even without power backoff, thereby ensuring PA efficiency. For example, both Gaussian frequency shift keying (GFSK) modulation used by BLE and Chirp modulation used by UWB ensure low-power consumption of wireless modules. However, for Wi-Fi networks that use the multicarrier modulation technology, PA efficiency is low due to the relatively high peak to average power ratio, greatly reducing the endurance.

2. Refresh rate

A larger amount of data sent by terminals or electronic tags in a unit time means higher power consumption, thereby reducing battery life. Therefore, for battery-powered tags, a trade-off between the battery life and refresh rate is needed. For Wi-Fi terminals or electronic tags, battery life is a weakness due to high power consumption. Therefore, Wi-Fi 6 offers enhanced energy-saving technologies to further improve the battery life. For example, the target wake-up time (TWT) technology determines the sleep time and active time of terminals or electronic tags, effectively shortening unnecessary active time periods. The Operating Mode Indication (OMI) technology uses low-power transmission in active time periods. Low-power terminals or electronic tags that support only 20 MHz bandwidth are compatible, and independent transmission schemes are designed for them.

6.3.2.3 Cost

- Equipment cost: the cost of equipment, including hardware and software, that must be purchased for building the positioning system.

- Deployment cost: the labor cost for deploying the positioning system, such as fingerprint collection and device installation position adjustment.

6.3.3 Comparison of Short-Range Wireless Positioning Solutions

Table 6.1 compares the common short-range wireless positioning solutions.

TABLE 6.1   Comparison of Short-Range Wireless Positioning Solutions

Positioning Solution	Precision (m)	Power Consumption	Deployment Cost
Wi-Fi RSSI-based positioning	3–7	High	Wi-Fi network construction cost
Wi-Fi RF pattern matching	3–5	High	Wi-Fi network construction cost and significant labor cost
Wi-Fi AoA-based positioning	1–3	High	High costs due to AP installation apart from Wi-Fi network construction costs
Wi-Fi FTM	1–2	High	Wi-Fi network construction cost
Bluetooth RSSI-based positioning	3–5	Low	Cost higher than that of Wi-Fi RSSI-based positioning due to slightly higher network deployment density than that of Wi-Fi
UWB	0.1–0.2	Low	High deployment costs due to dedicated devices and networks and strict network planning conditions

6.4 REQUIREMENT-BASED POSITIONING SYSTEM DESIGN

The previous sections discussed the principles of positioning technologies and the indoor positioning solutions backed by short-range wireless communications systems. The following sections will describe the composition and construction of positioning systems in regards to their requirements.

6.4.1 Introduction to Positioning Systems

A positioning system, regardless of the positioning technology and wireless standard used, consists of several fixed parts, as illustrated in Figure 6.14.

1. Data source

    The data source is the location of the original data, and it consists of two parts. The data source's first part is data that can be used for location calculation, such as wireless data, geomagnetic data, and motion data. That is, this data originates from either machines, natural environments, or human bodies. More, wireless data such as Wi-Fi signals is generated by APs or STAs, and such signals carry information to be used for location calculation, for example, signal strength, channel status, and time. In contrast, geomagnetic data is

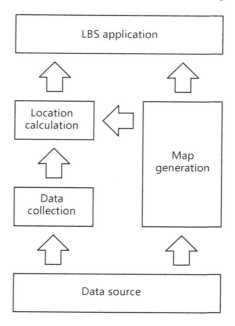

FIGURE 6.14   Positioning system.

sourced from the earth and is present in indoor locations, while it can also be used for location calculation. Lastly, motion data refers to information (such as acceleration) generated by a moving object, that is, an object's location can be calculated as long as the starting point and acceleration are known. The data source's second part is data that is used to draw maps, including building CAD drawings, PDF drawings, and building information models (BIMs) which are processed and organized to obtain easy-to-read maps.

2. Data collection

The data used for location calculation is collected from data sources, as illustrated in Figure 6.15. A data collection module receives and collects source data, and converts it into data that can be identified by the location calculation module, before sending the converted data to the location calculation module. An AP, for example, is a wireless switching device that must be able to receive and send Wi-Fi packets. It converts the data it receives into useable information, such as RSSI, timestamp, or signal AoA, according to the requirements of the location calculation module. However, the data conversion principles of an AP differ between data formats. For example, RSSI and timestamp conversion are both relatively simple processes, while signal AoA conversion is more complex. No matter

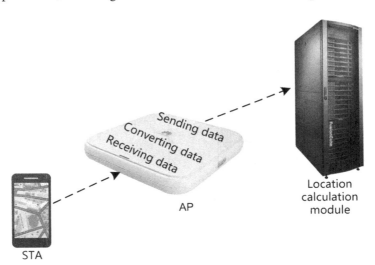

FIGURE 6.15   Data collection module.

what data format is chosen, the positioning system requires fast and accurate data conversion. After conversion, the data is sent to the location calculation module using a specified protocol format that should, at the very least, ensure high transmission efficiency, low resource consumption, and high levels of security.

What's more, it is worth noting that the data collection device may not necessarily be APs. STAs, for example, can also serve this function.

3. Location calculation

Location calculation is essential to a positioning system.

The location calculation module's main function is to calculate the coordinates of a located object based on the data received from the data collection module. The module also provides other functions such as open interfaces, path correction, and data collection device location marking. Based on different input data, the location calculation module uses different calculation mechanisms. For example, when the input data is RSSI, timestamp, and AoA, the location calculation module uses triangulation, positioning based on TDoA or time of flight (ToF), and AoA-based positioning, respectively. The location calculation module, sharing similarities with the data conversion module, also requires accurate and fast location calculation. The location calculation module's open interface function ensures that calculated location information can be used for other services through industry universal interface protocols (e.g., RESTful). However, as positioning services are only applicable when used in navigation or other professional applications, it is still yet to meet the application needs of different industries. The module's path correction function, also referred to as auxiliary positioning, is employed to minimize positioning errors. This is because in many cases, the data collection device sends multiple types of data used to correct STA locations, including geomagnetic data and motion data. In the path correction function, data is collected to establish a location feature database. Then, when the received data matches a location feature in the database, the coordinates are marked as object locations. Finally, the object's location is corrected using basic logic of path correction (e.g., an object cannot pass through a wall or cannot "walk" outside the mapped area). As such, well-functioning location calculation devices must rely on powerful path correction, and the data collection device location marking function provides a convenient and quick marking method. Without its function, a heavy workload is required to provide device positioning services.

4. Map generation

A map generation module has two basic functions. Its first function is to generate a map based on PDF drawings, CAD drawings, and BIM information. The map (which can be either 2D or 3D) is used to display a real environment. Its second function is path planning, a function that is often used in navigation. In path planning, after the user enters the source and destination, a navigation path is displayed on the map. The information generated by these two functions, when used together, is provided to the location calculation device and upper-layer applications through open interfaces.

5. Location based service (LBS)

The LBS module is used to manage people or objects in a real environment based on the location information of targets and the map where the targets are located. It is used, for example, in customer flow analysis, tracking or backtracking, and navigation. In customer flow analysis, crowd density in different areas and time segments can be visually displayed on the map. By contrast, in tracking and backtracking, terminal tracks are displayed on the map, and in navigation scenarios, real-time terminal locations are displayed on a map to help locate specific people.

The modules of a positioning system can belong to more than one physical device, while their positioning solutions can differ based on their application. In actual deployment, it is recommended to select effective positioning solutions based on customer requirements, which are classified into either network-side positioning or terminal-side positioning. Once customer requirements become clear, select a positioning solution based on the positioning precision, selected technology, and specific user group.

6.4.2 Network-Side Positioning Solutions

Network-side positioning requires that location information be presented on a network-side server. Then, customers obtain the location of a specific target without the target knowing its location information. To deliver network-side location, the required information needs to be obtained through the server by performing the following solutions: customer flow analysis, personnel management, and asset management.

The following describes the implementation of these different solutions.

1. Customer flow analysis

Used in densely populated areas, such as shopping malls, exhibition halls, airports, and sports stadiums, its purpose is to provide information such as the crowd distribution and store access volume, so that services can be improved. However, as the abovementioned scenarios are accessed by guests rather than regular visitors, the ideal solution involves using guests' smart terminals as the target objects for positioning.

To ensure meter-level precision, RSSI-based positioning that uses Wi-Fi or Bluetooth is deployed to locate smart terminals that also support Wi-Fi or Bluetooth. However, Wi-Fi is most commonly used, as it is most frequently found on customers' smart devices.

In this solution, the Wi-Fi packets sent from smart terminals are the data source, APs serve as the data collection device, while an independent location engine serves as the location calculation device, an independent map engine generates maps, and an LBS application server serves as the network-side server. The networking of the customer flow analysis solution is illustrated in Figure 6.16.

In this solution, the location engine, map engine, and business intelligence server are used by customers as software and installed on either a locally deployed server or cloud server. What's more, the AP's spacing, installation height, and antenna type must also meet the positioning network's requirements to be more effective. Table 6.2 lists the recommended installation parameters.

However, in actual deployment, the above conditions are often difficult to achieve. In any case, the auxiliary location correction function of the location engine is used to meet customers' standards. For example, if the linear deployment of APs in corridors does not meet the customers' requirements, a map is needed to adjust the target object location.

2. Asset management

In the asset management solution (outlined in Section 5.2), Wi-Fi or Bluetooth packets sent from electronic tags are the data source, while APs serve as the data collection device, an independent location

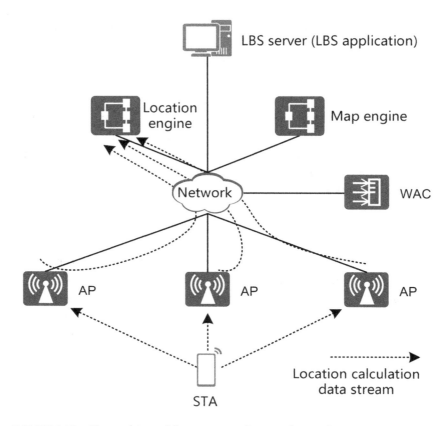

FIGURE 6.16   Networking of the customer flow analysis solution.

TABLE 6.2   Recommended Installation Parameters of APs for Customer Flow Analysis

Spacing	Installation Height	Antenna Selection	Installation Location
<20 m	<5 m	Omnidirectional antenna	Target area covered by at least three APs

engine serves as the location calculation device, an independent map engine generates maps, and the network-side server provides the LBS application.

3. High-precision positioning

   High-precision positioning is applicable in industrial scenarios where precision within 1 m is required. Such scenarios include a geofence that detects the terminal location to determine whether it crosses the geofence boundary, and the positioning system used in vehicle production lines that locates vehicles and determines whether vehicles are parked accurately. This solution is most applicable to scenarios that require high positioning precision and Wi-Fi coverage. Huawei IoT APs integrate the high-precision positioning modules of mainstream UWB vendors to converge two networks and provide indoor high-precision positioning for customers.

The UWB positioning system consists of the UWB base station, UWB tag, location engine, map engine, and LBS application. In this system, packets that are sent by the UWB tag are the data source, the UWB base station serves as the data collection device, the location engine serves as the location calculation device, an independent map engine generates maps, and the network-side server provides the LBS application. In the high-precision positioning converged solution, the UWB base station is integrated into an AP in the form of a card, providing both network services and high-precision positioning while also remaining cost-effective. Figure 6.17 illustrates the networking of the high-precision positioning solution.

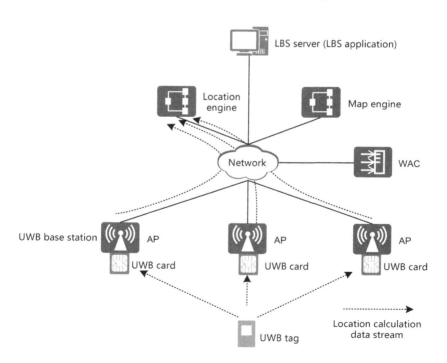

FIGURE 6.17   High-precision positioning solution networking.

To achieve sub-meter-level precision, the solution requires that strict deployment requirements are satisfied. For example, the location engine and APs must be deployed on the same WLAN, while no obstacles can obstruct the APs in the same location unit. Additionally, the distance between APs must be less than 50 m, and the AP's installation height needs to be less than 5 m.

6.4.3 Terminal-Side Positioning Solutions

Terminal-side positioning requires that location information be presented on terminals, while customers want the target to know its location without caring whether the location information is sent to a network-side server. Terminal-side positioning is mainly used to support navigation for users. Specifically, this solution needs to include the capability to display maps and location information on smart terminals in situations where users need to quickly and easily reach their destinations.

In this solution, meter-level precision must be supported, and the location refresh delay must be less than 3 seconds. In this case, terminal-side solutions are usually based on Bluetooth iBeacon. Figure 6.18 illustrates the networking of two common terminal-side location solutions. In both solutions, broadcast frames from the Bluetooth iBeacon are the data source, and a mobile phone serves as the data collection device and

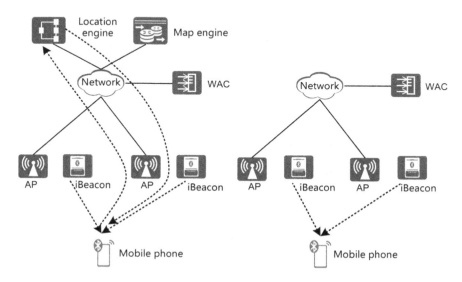

FIGURE 6.18 Terminal-side positioning networking.

provides the LBS application. The difference between the two, however, lies in the location calculation device and the map generation device.

The left side of Figure 6.18 illustrates the location calculation device and the map generation device as the location engine and the map engine, respectively. After collecting the Bluetooth iBeacon signals, the mobile phone sends the original location data to the location engine over the Wi-Fi network. Then, the location engine calculates the location and returns the information to the mobile phone. Meanwhile, the mobile phone continuously obtains map information about a new location from the map engine as the location changes. In this networking session, mobile phones use simple applets to enable navigation.

By contrast, on the right of Figure 6.18, the mobile phone serves as both the location calculation device and the map generation device. In general, mobile phones use independent apps to provide navigation services.

When deploying this solution, iBeacon base stations must be less than 8 m apart and installed at a height lower than 5 m. In addition, APs with built-in Bluetooth modules can also function as iBeacon base stations.

CHAPTER 7

Enterprise WLAN Networking Design

ENTERPRISES HAVE VARYING REQUIREMENTS for wireless local area networks (WLANs). As WLAN application becomes greater, it is becoming more important for enterprises to build WLANs that meet service requirements. Additionally, before building a robust WLAN, an effective architecture and networking mode must be designed. This chapter expands upon WLAN design ideas and typical networking solutions.

7.1 WLAN COMPOSITION

To facilitate your understanding of different networking architectures, it is necessary to learn the basic elements and components of a WLAN network, including how to connect and construct a wireless network between units.

7.1.1 Basic Units of a WLAN

Basic service set (BSS) is a WLAN's basic unit. It always consists of one AP along with multiple STAs where, as an infrastructure, the AP provides wireless communications services for STAs. This is illustrated in Figure 7.1. As such, the AP is the BSS center. Its location is fixed and it determines the BSS location. STAs are distributed around the AP and their locations are not fixed relative to the AP. They can move as close to or as away from the AP as specified. In addition, the AP coverage area, called the basic service area (BSA) or a cell, allows STAs to freely access or leave a cell, while only STAs that have accessed the cell can communicate with the AP.

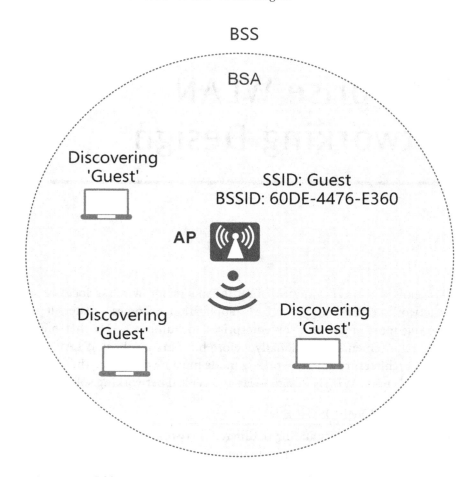

FIGURE 7.1   BSS.

Although STAs can communicate with the AP, multiple STAs in a cell cannot directly communicate with each other. If every STA sends packets randomly, severe conflicts and interference will occur on the cell channels. This can be better understood by comparing an AP to a traffic light and STAs to the vehicles driving on the road. Using this example, the vehicles are only allowed to pass through when the traffic light permits. If they were to ignore the traffic light, traffic accidents and congestion would occur. This is similar to how STAs in a cell can only communicate with the AP.

In the cell, the AP broadcasts its operating channel and coding and modulation schemes. It also manages wireless connections with STAs and sets up an order. This way, STAs can discover and locate the AP, share channel resources in order, and sequentially communicate with the AP. Communication data between STAs is forwarded by the AP.

1. BSS identity

For STAs to discover and locate an AP, the AP must have an identity and notify the STAs of its identity, also known as the basic service set identifier (BSSID). As the BSSID must be unique in order to differentiate BSSs, the MAC address of an AP is used.

However, if multiple BSSs are deployed in a space, an STA may discover multiple BSSIDs. In this case, only the BSSID of the BSS to which the STA is to be added needs to be selected. As STAs do not automatically select the BSSIDs, the choice is up to the user. However, the user may be illustrated a series of MAC addresses on an STA, making it harder for them to select only one. To avoid this confusion, a freely configurable character string, also known as a SSID, is required to make up the AP name and facilitate AP identification. It is worth noting that although the SSID and BSSID are different, the same SSID can be configured for different BSSs. That is, if the BSSID is the BSS ID card, the SSID is the BSS name. The WLAN name found on an STA is the SSID.

2. Coexistence of multiple BSSs

When first deployed, only one BSS was able to be configured for an AP. And if multiple BSSs were required in the same space, multiple APs needed to be deployed. By doing this, not only costs were increased, but channel planning between different BSSs became extremely complex. What's more, by doing this, channel resources became even more insufficient. However, after various improvements, most APs can now be engineered into multiple virtual access points (VAPs) apiece, with each encompassing only one BSS. This ensures that multiple BSSs can be deployed with only one AP. With different SSIDs configured for these BSSs, users will notice multiple WLAN networks on an STA, or alternatively multiple SSIDs.

In addition, as the BSSID is an AP's MAC address, an equivalent number of MAC addresses need to be configured based on the number of VAPs supported by the AP. The relationship between VAPs, SSIDs, and BSSIDs is illustrated in Figure 7.2.

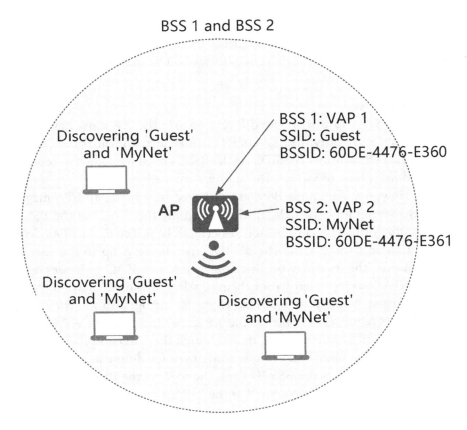

FIGURE 7.2   Coexistence of multiple BSSs.

Although VAPs simplify WLAN deployment, they do not necessarily deliver improved performance. This is because increasing VAPs
not only require users to spend more time finding SSIDs, but AP
configuration also becomes more complex. Therefore, the quantity of
VAPs used must be planned based on actual requirements. As VAPs
share the software and hardware resources of one AP, while VAP
users share channel resources, the AP capacity remains unchanged
and does not multiply with the number of VAPs.

7.1.2 Distribution System

BSSs can deliver wireless communication between multiple STAs in a specified area. However, as the objects with which STAs communicate can be
scattered in different areas, APs need to be connected to a larger network

to link up the BSSs in different areas. This network is the upstream network of an AP, also known as the distribution system (DS) of the BSS, as illustrated in Figure 7.3.

As the upstream network of an AP is usually an Ethernet network, the AP can connect to the upstream network over radios or wired interfaces. Upon receiving wireless packets from an STA, an AP converts them into wired packets before sending them to the upstream network. Then, the upstream network forwards the packets to another AP. The upstream network of an AP can also be a wireless network. For example, in areas where cabling is difficult, an AP can interact with other APs working in bridge

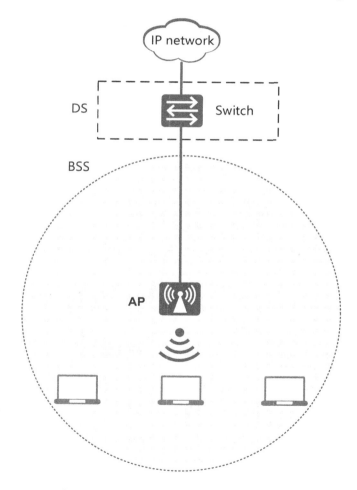

FIGURE 7.3   DS of the BSS.

mode through wireless connections, or it can be configured with LTE functions to access mobile networks.

However, the effective coverage radius of a BSS, generally 10–15 m, does not provide enough support for large buildings or exhibition halls. To address this challenge, more BSSs can be deployed, while APs can be arranged in an evenly spaced manner and connected using the BSS's DS. By doing so, users are afforded access to WLAN networks from anywhere.

Another challenge that needs to be addressed is the inconvenience caused by users needing to search for an SSID every time they move from one BSS to another to join in the another BSS. Throughout the process, users may also encounter network disconnections. Therefore, to eliminate this impact, each BSS is able to use the same SSID, so that users experience the same WLAN network wherever they go.

This method of extending the BSS range is also known as extended service set (ESS), and it is a flexible combination of BSSs to enable more flexible WLAN deployment, as illustrated in Figure 7.4. In addition, the

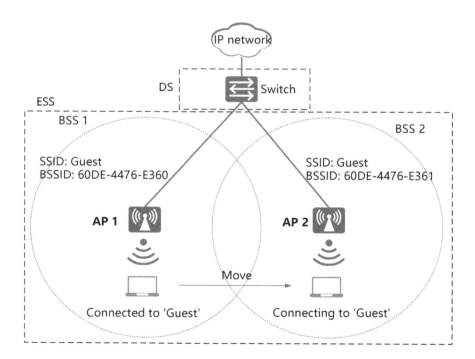

FIGURE 7.4   ESS.

same SSID for all BSSs is used as the ESS identifier (ESSID), which is broadcast to STAs to notify them of a roaming WLAN.

The network administrator manages not only user terminal movement and roaming within an ESS, but also the APs that set up an ESS. It also manages terminal access, terminal roaming efficiency, and network security, and ensures that network problems are located quickly and networks remain optimized. As such, the selection of a WLAN networking architecture holds great importance.

7.2 WLAN NETWORKING ARCHITECTURE

In the beginning of this book, the evolution of the WLAN network architecture from fat AP to WAC+Fit AP was described. This section discusses the differences between these two networking architectures and the factors that affect the WAC+Fit AP architecture. It will also introduce three new networking architectures: cloud management architecture for small- and medium-sized enterprises, leader AP architecture for small and micro enterprises, and agile distributed architecture for room-intensive scenarios. Finally, it will introduce the next-generation intent-driven campus network architecture that aligns with the digital transformation of industries. These networking architectures are compared in Table 7.1.

7.2.1 Fat AP Architecture

The fat AP architecture, also called the autonomous network architecture, implements wireless access, service data encryption, and service packet forwarding without centralized control by a dedicated device. Figure 7.5 illustrates this architecture.

However, in the fat AP architecture, autonomy has both its benefits and disadvantages. Fat APs are ideal for home WLANs due to their low costs and simple deployment, independent operation, and no need for centralized control. However, the downside of fat APs is that they are not suitable for large-scale enterprises. That is, as the WLAN coverage area expands and users that access the WLAN increase, more fat APs need to be deployed. As these APs work independently and there is no unified control device, their management and maintenance become challenging for network managers. For example, software upgrade needs to be performed independently for each fat AP, a process that is both time-consuming and labor-intensive. In addition, this architecture cannot support user roaming in a larger coverage area, and the unified control of complex services,

TABLE 7.1   WLAN Networking Architecture Comparison

Networking Architecture	Scope of Application	Characteristics
Fat AP	Home	The AP is independently configured and inexpensive and provides simple functions
WAC+Fit AP	Large- and medium-sized enterprises	APs are managed and configured by the WAC and have abundant functions. The skill requirements of maintenance personnel are high
Cloud management	Small- and medium-sized enterprises	APs are managed and configured by the cloud management platform, have abundant functions, and support plug-and-play. The skill requirements of maintenance personnel are low
Leader AP	Small and micro enterprises	The leader AP can work independently or manage a few APs to implement basic roaming functions. The cost and skill requirements are low
Agile distributed architecture	Scenarios with densely distributed rooms	An AP with a special architecture is divided into a central AP and a RU The central AP can manage multiple RUs. It features low costs and good coverage Agile distributed APs can be used in any of the fat AP, WAC+Fit AP, and cloud management architectures
Intent-driven campus network	Large- and medium-sized enterprises	APs are managed and configured by the SDN controller and provide abundant functions. They can further integrate with wired networks and work with big data and AI technologies to implement simplified, smart, and secure campus networks

for example, priority policy control based on different data types of network users.

Therefore, the fat AP architecture is less suitable for enterprises than other architectures such as WAC+Fit AP, cloud management, and leader AP.

7.2.2 WAC+Fit AP Architecture

Figure 7.6 illustrates the WAC+Fit AP architecture. In this architecture, the WAC is responsible for WLAN access control, forwarding, statistics collection, AP configuration monitoring, roaming management, AP network management agent, and security control. The fit AP, which is

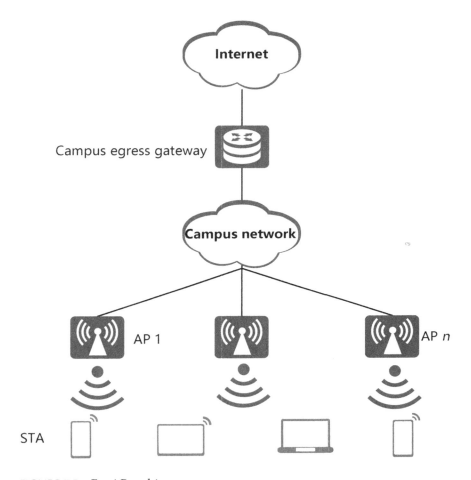

FIGURE 7.5   Fat AP architecture.

managed by the WAC, encrypts and decrypts 802.11 packets, transmits them at the physical layer, and collects statistics on the air interface.

The WAC communicates with APs through the Control and Provisioning of Wireless Access Points (CAPWAP) protocol. This protocol defines the following contents: automatic WAC discovery by APs, security authentication of APs by the WAC, software acquisition by APs from the WAC, and initial and dynamic configuration acquisition by APs from the WAC. In communication, an AP and the WAC establish a CAPWAP tunnel, which may be a control tunnel or a data tunnel. The control tunnel carries control packets (or management packets for the WAC to manage APs),

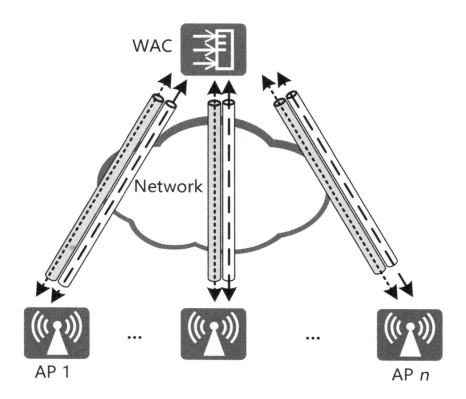

FIGURE 7.6   WAC+Fit AP architecture.

while the data tunnel carries data packets. CAPWAP tunnels improve packet transmission security by using Datagram Transport Layer Security (DTLS) encryption.

Compared with the fat AP architecture, the WAC+Fit AP architecture has the following advantages:

- Configuration and deployment: The WAC implements centralized network configuration and management, which can automatically adjust the channels and power of APs in a network.

- Security: In the fat AP architecture, APs cannot be upgraded at the same time. Therefore, some APs may not have the latest software patches. In the WAC+Fit AP architecture, however, the WAC provides most security functions and implements software updates and security configurations, facilitating global security settings. Additionally, to prevent malicious codes from being loaded, devices perform digital signature authentication on the software, further enhancing update security.

- The WAC also provides some security functions that are unavailable in the fat AP architecture. These include advanced security features such as virus detection, uniform resource locator (URL) filtering, and stateful inspection firewall.

- Update and expansion: Centralized management of the WAC+Fit AP architecture enables APs under the same WAC to share a software version. When an AP needs to update, it obtains the update package from the WAC, which then performs the update on the AP. Updating is simple because the functions of the AP and WAC are separated, with the WAC performing only user authentication, network management, and security functions.

Networking and data forwarding modes, along with multiple WACs in the WAC+Fit AP architecture, affect WLAN quality. The following sections describe these factors in greater detail.

7.2.2.1 Networking Mode

The network between a WAC and fit APs can be a Layer 2 or Layer 3 network, as illustrated in Figure 7.7.

- In Layer 2 networking, the WAC and fit APs are in the same broadcast domain. This networking mode is simple, as APs can directly discover the WAC through local broadcast, but it is not suitable for large-scale networks.

- In Layer 3 networking, the WAC and fit APs are in different network segments, which complicates configuration. The intermediate network must ensure that the routes between the APs and WAC are reachable. Extra configurations are required so that the APs can

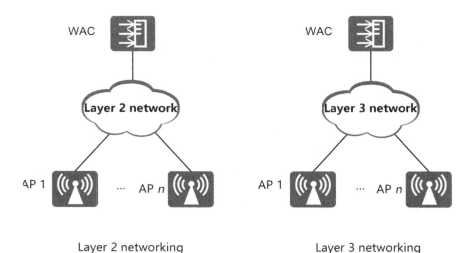

Layer 2 networking   Layer 3 networking

FIGURE 7.7   Layer 2 and Layer 3 networking.

discover the WAC. Layer 3 networking is applicable to medium- and large-sized networks. For example, in a large campus, APs are deployed in each building to provide wireless coverage, while the WAC is deployed in the core equipment room to centrally manage all APs. This deployment necessitates a complex Layer 3 network between the WAC and fit APs.

A WAC can be deployed in inline or bypass networking mode, as illustrated in Figure 7.8.

1. Inline networking

    In this mode, the WAC also functions as an aggregation switch to process and forward data and management traffic of APs. Inline networking is suitable for new small- and medium-scale WLANs that are deployed in a simple, centralized manner.

2. Bypass networking

    The bypass mode meets the high WAC requirements in the inline mode. The WAC is connected to a live network (usually beside the aggregation switch) and only manages APs. In this way, data services can be directly transmitted to the upstream network without passing through the aggregation switch. Bypass networking can also be used

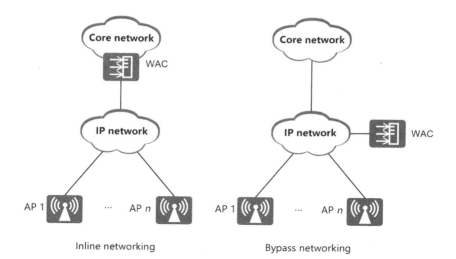

FIGURE 7.8   Inline and bypass networking.

for live network reconstruction. For example, if a WLAN is required on a wired live network, a WAC can be effortlessly connected to the aggregation switch. This mode is suitable for large- and medium-scale WLANs.

7.2.2.2 Data Forwarding Mode
Data packets can be forwarded in direct or tunnel mode.

1. Direct forwarding

    Direct forwarding is also referred to as local data forwarding. Data packets are directly forwarded to the upper-layer network without being encapsulated in the CAPWAP tunnel by APs, as illustrated in Figure 7.9.

    Direct forwarding is often used in inline networking. Direct forwarding in inline mode simplifies the network architecture and applies to medium- and small-scale, centralized WLANs.

    Direct forwarding can also be used in bypass networking mode. Data packets do not need to be processed by the WAC in a centralized manner, eliminating bandwidth bottlenecks, while the security policies of live networks can be retained. This mode is applicable to large-scale campuses with integrated wired and wireless networks or branches.

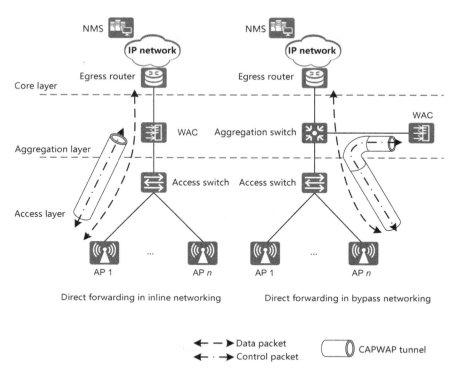

FIGURE 7.9 Direct forwarding mode.

2. Tunnel forwarding

In tunnel forwarding, which is also referred to as centralized data forwarding, data packets are encapsulated in a CAPWAP tunnel and then forwarded by the WAC to the upper-layer network, as illustrated in Figure 7.10. Tunnel forwarding is usually used together with bypass networking. The WAC forwards data packets, ensuring high security, facilitating centralized management and control, simplifying deploying, and configuring new devices. All in all, it requires a little live network reconstruction. This mode applies to large-scale campuses with WLANs deployed independently or centrally managed and controlled.

7.2.2.3 VLAN Planning

VLANs on a WLAN are broken into management VLANs and service VLANs. A management VLAN transmits packets forwarded through the CAPWAP tunnel, including management packets and CAPWAP

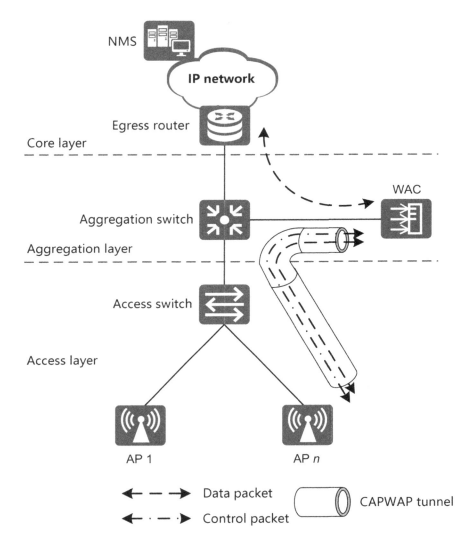

FIGURE 7.10   Tunnel forwarding mode.

tunnel-forwarded service data packets. Conversely, a service VLAN transmits only service data packets.

VLAN planning should ensure:

- Management VLANs are isolated from service VLANs.

- Service VLANs and SSIDs are mapped based on service requirements.

TABLE 7.2   Mapping between VLANs and SSIDs

Mapping Relationship	Example
1:1	An enterprise needs to deploy a WLAN in both areas A and B, but wants users to use the same SSID and same data forwarding control policy. In this case, only one SSID and one VLAN need to be planned. As a result, the SSID:VLAN mapping relationship is 1:1
1:N	An enterprise needs to deploy a WLAN in both areas A and B, but wants users to use the same SSID and different data forwarding control policies. In this case, one SSID and two VLANs (corresponding to different areas) are needed. As a result, the SSID:VLAN mapping relationship is 1:2
N:1	An enterprise needs to deploy a WLAN in both areas A and B, but wants users to be provided location information after they detect a WLAN while using the same data forwarding control policy. In this case, two SSIDs (SSID_A for area A and SSID_B for area B) and one VLAN can be planned. As a result, the SSID:VLAN mapping relationship is 2:1
N:M	An enterprise needs to deploy a WLAN in both areas A and B, but wants users to be provided location information after they detect a WLAN while using different data forwarding control policies. In this case, two SSIDs (SSID_A for area A and SSID_B for area B) and two VLANs (VLAN_A for area A and VLAN_B for area B) can be planned. As a result, the SSID:VLAN mapping relationship is 2:2

Both service VLANs and WLAN SSIDs can identify services and users. For this reason, the mapping between VLANs and SSIDs must be considered during WLAN planning. Four mapping relationships are available based on scenarios: 1:1, 1:N, N:1, and N:M. For details, see Table 7.2.

A special WLAN scenario is that a large number of users often roam from their first accessed networks to other areas. As a result, the first accessed networks require a large number of IP addresses, which is a common phenomenon at stadium entrances and hotel lobbies. If one SSID corresponds to only one VLAN and one VLAN corresponds to one subnet, when a large number of users in an area access the network, the subnet of the VLAN corresponding to the SSID must expand to ensure that users can obtain IP addresses. As a result, the broadcast domain expands and a large number of packets, such as Address Resolution Protocol (ARP) and Dynamic Host Configuration Protocol (DHCP) packets are broadcast, causing severe network congestion. To address this issue, a VLAN pool, which has multiple VLAN management and allocation algorithms, can be used as a service VLAN, so that one SSID corresponds to multiple

VLANs. This way, a large number of users are distributed to different VLANs to narrow the broadcast domain.

For example, in a stadium (as illustrated in Figure 7.11), multiple WACs are required to provide network access for a large number of users. However, users may frequently move around the stadium, roaming between APs or even across WACs. In this case, the number of VLANs required in a VLAN pool can be evaluated based on the number of users accessing the WLAN in the stadium. The SSID can be bound to the VLAN pool so that users can access different VLANs, narrowing the broadcast domain. Moreover, after a VLAN pool has been bound, users in the same VLAN pool can roam on the Layer 2 network but not on the Layer 3 network, improving WLAN performance.

7.2.2.4 IP Address Planning

IP address planning should be based on the quantities of WACs, APs, and STAs in a network.

- WAC IP address: used for managing APs. In most cases, WAC IP addresses are limited and they are configured statically and manually.

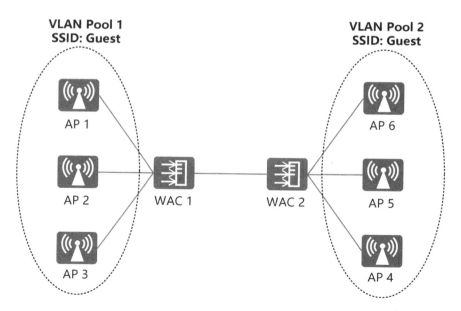

FIGURE 7.11   VLAN pool application example.

- AP IP address: used for communication with the WAC through a CAPWAP tunnel. Due to a large number of APs, an independent DHCP server, or preferentially a WAC that also functions as a DHCP server, dynamically allocates IP addresses. If the WAC or DHCP server is connected to APs over a Layer 3 network, a DHCP relay must be configured on the intermediate network, and the route between the WAC or DHCP server and the DHCP relay must be reachable.

- STA IP address: It is recommended that an independent DHCP server or a WAC that also functions as a DHCP server dynamically allocates STA IP addresses. Static configuration is not recommended, except for stationary wireless STAs (such as wireless printers).

7.2.2.5 Comparison of Typical WAC+Fit AP Networking Modes
Table 7.3 examines and compares four WAC+Fit AP networking modes.

7.2.2.6 WAC Backup
In network design, redundant devices and links, as well as deployment switchover policies, should be taken into account to ensure the system functions are not affected by single points of failure. WAC backup design is a key aspect of the WAC+Fit AP architecture.

TABLE 7.3   Comparison between Typical WAC+Fit AP Networking Modes

Networking Mode	Pros	Cons
Inline + Layer 2 + direct forwarding	Data traffic is forwarded without roundabouts	Complex data VLAN configurations
Bypass + Layer 2 + direct forwarding	Data traffic is forwarded without roundabouts. WLAN deployment on legacy networks and hot backup are simplified	Complex data VLAN configurations
Bypass + Layer 2 + tunnel forwarding	Data VLAN configurations are effortless. Layer 2 tunnel forwarding and 802.1X authentication are supported. WLAN deployment on legacy networks and hot backup are simplified	Forwarding efficiency is low and data traffic is detoured
Bypass + Layer 3 + tunnel forwarding	Data VLAN configurations are effortless. Layer 2 tunnel forwarding and 802.1X authentication are supported. WLAN deployment on legacy networks and hot backup are simplified	Forwarding efficiency is low and data traffic is detoured

There are three common WAC backup solutions: Virtual Router Redundancy Protocol (VRRP) based hot standby (HSB), dual-link HSB, and $N + 1$ backup.

1. VRRP HSB

In VRRP HSB mode, two WACs form a VRRP group, whereby the active and standby WACs share the same virtual IP address. The active WAC synchronizes service information to the standby WAC through the HSB channel, as illustrated in Figure 7.12. In this solution, APs detect only one WAC, and the VRRP determines the switchover between WACs. This mode enables faster service switchover compared to other modes because the active and standby WACs are deployed at the same location.

2. Dual-link HSB

An AP establishes CAPWAP tunnels with the active and standby WACs, and the HSB channel synchronizes service data between the

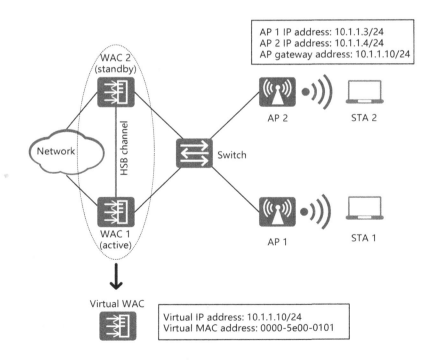

FIGURE 7.12   VRRP HSB networking.

WACs. When the link between the AP and the active WAC is disconnected, the AP notifies the standby WAC to become the active WAC.

Dual-link HSB supports both active/standby backup and load balancing. As Figure 7.13 illustrates, in load balancing mode, WAC 1 is the active WAC of AP 2 and establishes a primary CAPWAP link to AP 2. Meanwhile, WAC 2 is the active WAC of AP 1 and establishes a primary CAPWAP link to AP 1. If WAC 1 experiences a fault, WAC 2 replaces WAC 1, then manages AP 2 and establishes a CAPWAP link to AP 2 to provide services. WAC 2 continues to manage the service data of AP 1 on the same path.

In this mode, the active and standby WACs are not subject to location restrictions and can be flexibly deployed. They can also implement load balancing to improve resource utilization, but service switchover is slow.

3. $N + 1$ backup

In $N + 1$ backup, one WAC functions as the standby WAC to back up services for multiple active WACs on a WAC+Fit AP network, as illustrated in Figure 7.14. When the network is running normally, an AP sets up a CAPWAP link only with its active WAC. When the active WAC or CAPWAP link becomes faulty, the standby WAC takes over the AP management and establishes a CAPWAP link with the AP to provide services.

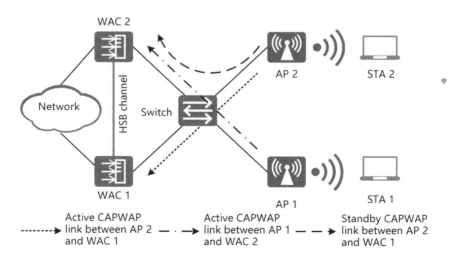

FIGURE 7.13   Dual-link HSB.

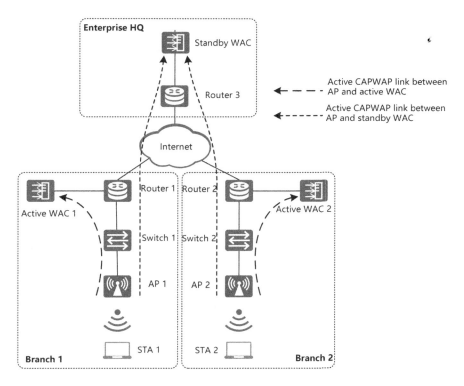

FIGURE 7.14   *N* + 1 backup.

Table 7.4 lists the differences between VRRP HSB, dual-link HSB, and *N* + 1 backup.

7.2.2.7 Inter-WAC Roaming

When multiple WACs need to be deployed on a large-scale network, they need to synchronize or query the status of roaming STAs in real time. This way, STAs can smoothly roam across WACs, ensuring the proper forwarding of traffic, implementing wireless coverage, and roaming in a wider range. As a solution to this, inter-WAC roaming is developed.

The following describes some basic concepts of inter-WAC roaming.

1. Roles involved in inter-WAC roaming

   • Home WAC (HWAC): a WAC with which an STA associates for the first time.

TABLE 7.4   Comparison between the Three Backup Solutions

Item[a]	VRRP HSB	Dual-Link HSB	N + 1 Backup
Switchover speed	The active/standby switchover is fast and has little impact on services. The VRRP preemption time can be configured to implement a faster switchover compared to other backup modes	The AP status switchover is slow and performed only after the CAPWAP link disconnection timeout is detected. After active/standby switchover, STAs do not need to reconnect	The AP status switchover is slow and performed only after the CAPWAP link disconnection timeout is detected. Both APs and STAs need to reconnect, causing short interruptions
Active/standby WAC remote deployment	VRRP is a Layer 2 protocol and does not support the remote deployment of active and standby WACs	Supported	Supported
Application scope	For scenarios that require high reliability but do not require remote deployment of active and standby WACs	For scenarios that require high reliability and remote deployment of active and standby WACs	For scenarios that have low reliability and are expensive

- Home AP (HAP): an AP with which an STA associates for the first time.

- Foreign WAC (FWAC): a WAC with which an STA associates after roaming.

- Foreign AP (FAP): an AP with which an STA associates after roaming.

An STA may roam multiple times, but the HWAC and HAP will not change, while the FWAC and FAP migrate continuously as each STA roams.

2. Inter-WAC tunnel

To support inter-WAC roaming, WACs need to exchange user information and forward user traffic. Therefore, a tunnel is established between WACs to transmit management and data packets.

3. Roaming group

Two WACs in a network may not necessarily support inter-WAC roaming. Therefore, a WAC group needs to be manually defined, permitting only WACs in the same group to roam between each other. This group is called a roaming group.

4. Layer 2 and Layer 3 roaming

Large-scale WLANs (e.g., a large campus network) will be accessed by a large number of STAs. The network administrator usually divides a WLAN into multiple subnets to reduce broadcast packets on the network. In a WLAN, Layer 2 roaming refers to STA roaming in the same subnet, while Layer 3 roaming refers to STA roaming between different subnets.

For VLAN pools, in particular, STA roaming in the same VLAN pool is Layer 2 roaming, while STA roaming in different VLAN pools is Layer 3 roaming.

Inter-WAC roaming is implemented as follows:

- In Layer 2 roaming, STAs remain in the same subnet before and after roaming. The FAP or FWAC forwards Layer 2 roaming STA packets in the same way as forwarding new online STA packets. The packets are forwarded on the local network of the FAP or FWAC, and do not need to be forwarded back to the HAP or HWAC over the inter-WAC tunnel.

- In Layer 3 roaming, STAs stay in different subnets before and after roaming. To enable the STAs to access the original network after roaming, service traffic must be forwarded to the original subnet over a tunnel. Specifically, a device that can communicate with the gateway on the original network at Layer 2 is selected as the home agent. Service traffic is then forwarded from the FAP to the home agent through a tunnel, and then forwarded onward by the home agent. Similarly, packets from the network to STAs are first sent to the home agent, and then forwarded to the FAP through the tunnel by the home agent.

The HAP or HWAC can both function as the home agent. Whether it should be a HAP or a HWAC depends on the data-forwarding mode:

- In tunnel forwarding mode, the HWAC acts as the home agent because the HWAC allows access to the user gateway. The tunnel between the FAP and the home agent consists of two segments: the inter-WAC tunnel between the FAP and the FWAC, as well as the CAPWAP tunnel between the FWAC and the HWAC. Data packets are forwarded from the FAP to the HWAC, and then forwarded to the upper-layer network through the HWAC, as illustrated in Figure 7.15.

- In direct forwarding mode, the HAP is the home agent by default. In this case, the tunnel between the FAP and the home agent

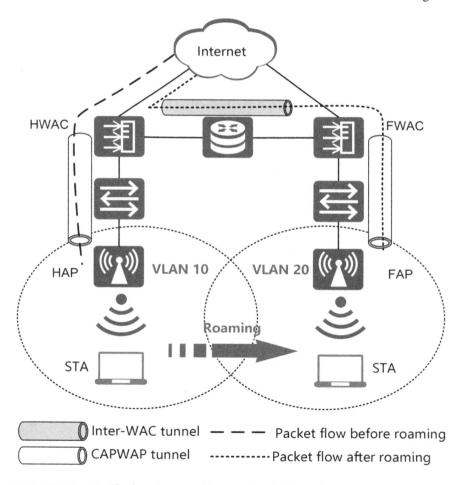

FIGURE 7.15   Traffic flow in tunnel forwarding in Layer 3 roaming.

consists of three segments: the CAPWAP tunnel between the FAP and the FWAC, the inter-WAC tunnel between the FWAC and the HWAC, and the CAPWAP tunnel between the HWAC and the HAP. Data packets are forwarded from the FAP to the HAP, and then forwarded by the HAP to the upper-layer network, as illustrated in Figure 7.16.

Particularly, if the HAP and the HWAC are in the same subnet, the most suitable HWAC will act as the home agent. This

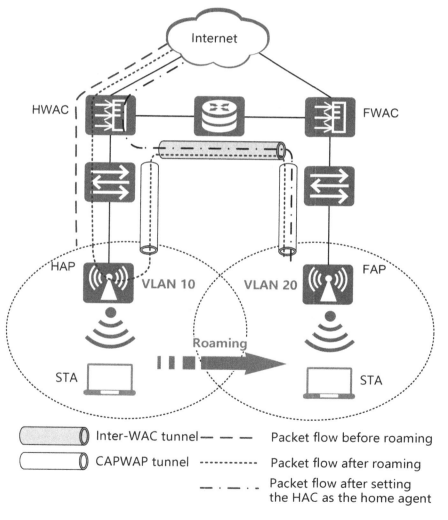

FIGURE 7.16   Traffic flow in direct forwarding in Layer 3 roaming.

way, traffic only needs to be forwarded to the HWAC, reducing the HAP load and improving forwarding efficiency.

7.2.2.8 Navi WAC

A large enterprise's wireless network needs to provide wireless access for both employees and guests, but guest data may bring potential security threats to the network. In Navi WAC networking, enterprises can divert guest traffic to the Navi WAC in the secure demilitarized zone (DMZ) for centralized management. This way, internal employee access is isolated from guest access.

As illustrated in Figure 7.17, employee traffic is forwarded on the enterprise intranet, and the employees can access the enterprise intranet server. Guest traffic is forwarded to the secure DMZ through the CAPWAP tunnel. In the DMZ, IP addresses are assigned to guests and authenticated. Guests can access only servers and Internet resources in the DMZ.

7.2.3 Cloud Management Architecture

Traditional network solutions suffer from high deployment costs and complex O&M, especially for enterprises with a large number of branches

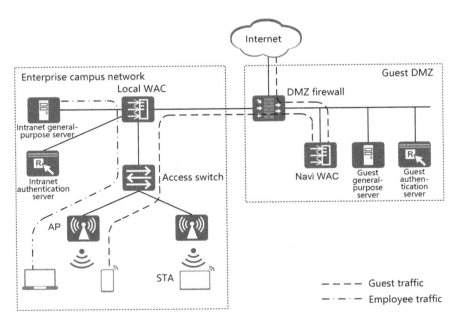

FIGURE 7.17 Typical Navi WAC networking.

or dispersed sites. As a solution to this, the cloud management architecture can centrally manage and maintain devices anywhere, greatly reducing O&M costs. Figure 7.18 illustrates the solution architecture.

After cloud APs are deployed, the network administrator does not need to commission them onsite. Once powered on, they automatically connect to the specified cloud management platform to load system files such as the configuration file, software package, and patch file. In this way, cloud APs can effortlessly go online. The network administrator can configure APs through the cloud management platform anytime and anywhere, facilitating batch service configuration.

Compared with the traditional WAC+Fit AP architecture, the cloud management architecture has the following advantages:

- Plug-and-play and automatic deployment reduce network costs.

- Unified O&M. All cloud-managed NEs are monitored and managed on the cloud management platform.

- Tool-based. In most cases, the cloud solution provides various cloud tools to effectively reduce OPEX. For example, Huawei CloudCampus solution provides end-to-end cloud tools (such as CloudCampus App).

7.2.4 Leader AP Architecture

Some small and micro enterprises aim at building and managing their own wireless networks without using a cloud management architecture. If the fat AP architecture is used, APs are not managed and maintained in a unified manner, and therefore cannot deliver an ideal roaming experience. If the WAC+Fit AP architecture is used, enterprises need to incur higher costs for WAC devices and certificates even though only a small number of APs are needed to completely cover a small area with a limited number of STAs. An AP that can manage other APs and provide unified O&M and continuous roaming functions is referred to as a leader AP, which meet the requirements of small and micro enterprises. Figure 7.19 illustrates the leader AP architecture.

The leader AP architecture consists of only APs. One of them is set as the leader AP, and other APs work in the fit AP mode and communicate with the leader AP through Layer 2. The leader AP then broadcasts its role on the Layer 2 network for other APs to automatically discover and connect. The functions of the leader AP are similar to those of the WAC,

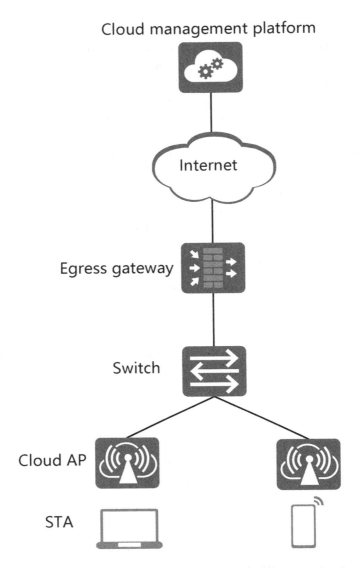

FIGURE 7.18 Cloud management architecture. (*Note*: Huawei cloud management platform: As a core component of Huawei CloudCampus Solution, iMaster NCE-Campus centrally manages Huawei network devices, such as APs, routers, switches, and firewalls. The cloud management platform not only implements functions such as unified multitenant management, plug-and-play network devices, and batch deployment of network services, it can also connect to a third-party platform by using an application programming interface (API) provided by the cloud management platform, to add more value-added services.)

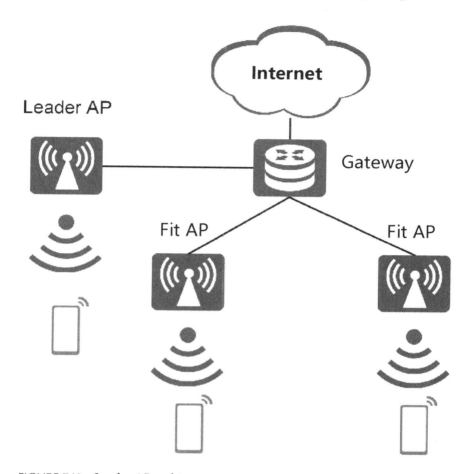

FIGURE 7.19   Leader AP architecture.

including CAPWAP tunnel-based unified access management, configuration and O&M, centralized radio resource management, and roaming management. Users only need to log in to the leader AP to configure wireless services. With this configuration, all APs provide the same wireless services, and STAs can roam between different APs.

7.2.5 Agile Distributed Architecture

In scenarios with densely distributed rooms — such as dormitories, hotels, and wards — and if the WAC+Fit AP architecture is used with one AP deployed in each room, numerous packets need to be sent to the WAC, posing high requirements on WAC performance. To resolve the WAC performance bottleneck and meet coverage requirements of each room, APs can be deployed in corridors with remote antennas to provide signals in every room. However, this solution has distance requirements, meaning longer distances create greater signal attenuation. In addition, signal quality is poor and user bandwidth is insufficient when multiple rooms share a single AP.

To solve this problem, Huawei has developed an innovative agile distributed architecture (see Figure 7.20) based on the traditional network architecture. The agile distributed architecture provides high bandwidth and eliminates coverage holes for a single room.

The agile distributed architecture divides traditional APs into central APs and remote units (RUs).

Central APs can be deployed in equipment rooms, weak-current wells, and corridors while RUs can be installed in rooms by routing cables through walls, allowing each room to have exclusive access to high-quality wireless services.

The agile distributed architecture redistributes and deploys service models on WACs, central APs, and RUs, with the following advantages:

- Simplified management: Only a few central APs need to be managed. Nearly 10,000 rooms require only 200 APs.

- Flexible deployment and full coverage: Central APs and RUs are deployed indoors through network cables without wall penetration attenuation and feeder loss, ensuring full signal coverage. An RU can be installed on a wall, ceiling, or panel.

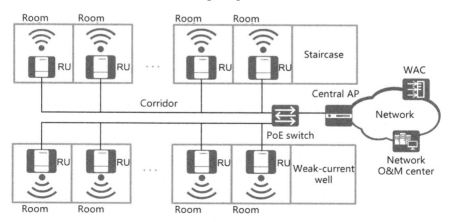

FIGURE 7.20   Agile distributed architecture.

- Ultra-long-distance coverage: Compared with traditional APs that support a distance of up to 15 m from antennas, a central AP can be deployed up to 100 m away from an RU, significantly expanding the network coverage range.

In hospitals, medical handheld terminals do not support 802.11k, 802.11v, and 802.11r standards. When medical workers use terminals for ward rounds, infusion check, and vital sign recording, roaming performance of the terminals is poor with packet loss or long delays. In this situation, the medical workers need to relogin to the application or rescan the barcode, which interrupts data transmission and severely hampers work efficiency.

Huawei's agile distributed single frequency network (SFN) roaming function is created to precisely address these problems. On an agile distributed WLAN, all RUs connected to a central AP are deployed on the same working channel and use the common BSSID for STAs to associate. When STAs move within the coverage area of the same SSID, the STAs are seamlessly connected and services are not interrupted. Compared with traditional intra-AP roaming, agile distributed SFN roaming does not result in varying roaming experience among different terminals. In addition, user reassociation, authentication, and key negotiation are not required during roaming, creating a fast and seamless experience with a significantly lower packet loss rate.

7.2.6 Development of Next-Generation Networking Architecture

Emerging technologies, such as the Internet of Things (IoT), big data, cloud computing, and artificial intelligence (AI), are set to transform enterprises' operations and production across industries. As a bridge connecting the physical and digital worlds, enterprise campus networks will inevitably face new challenges. To cope with the massive growth of data and applications, campus networks must be deployed more easily and quickly, run more securely and reliably, and be more intelligent in management and O&M. Huawei's next-generation campus network architecture — the intent-driven campus network — is designed to make it a reality for all enterprises. The core concepts of the intent-driven campus network architecture include ultra-broadband, simplicity, intelligence, security, and openness. Huawei hopes to provide the development direction of the future campus network for network professionals worldwide.

Figure 7.21 illustrates a typical intent-driven campus network architecture, which consists of the network, management, and application layers.

7.2.6.1 Network Layer

The network layer provides ultra-broadband capabilities. The intent-driven campus network uses virtualization technology to divide the network layer into the physical network (underlay network) and the virtual network (overlay network).

- Physical network: It provides basic connection services for campus networks. Similar to traditional campus networks, it can be divided into the access layer, aggregation layer, core layer, and egress zone. A physical network consists of switches, routers, firewalls, WACs, and APs.

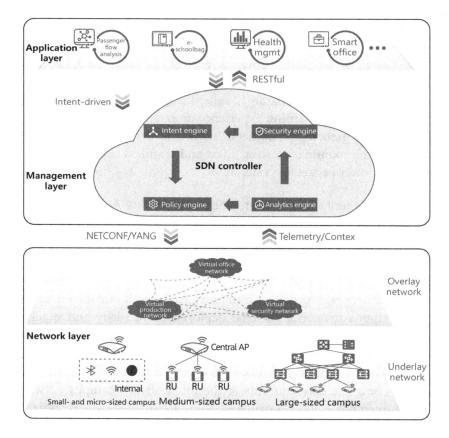

FIGURE 7.21   Intent-driven campus network architecture.

• Virtual network: One or more virtual networks are built on the physical network. Service policies are deployed on the virtual network to serve different services or customer groups.

The physical network is completely decoupled from the virtual network and it keeps evolving — according to Moore's Law — to build ultra-broadband forwarding and ultra-broadband access network capabilities. Virtual networks use the overlay technology to shield complex physical network devices. Then, the overlay technology is used to create reachable campus switching networks.

7.2.6.2 Management Layer

The management layer — the center of intelligence — provides network-level management services such as configuration management, service management, network maintenance, fault detection, and security threat analysis for campus networks. The intent-driven campus network is centered on the software-defined networking (SDN) controller. This SDN controller adopts the SDN management concept — in which service management is not device-specific — to extract the intent engine, policy engine, security engine, and analytics engine in campus network management. This concept is from a user and business perspective, see Table 7.5 for more information. The four engines integrate big data and AI technologies to build a simplified, intelligent, and secure campus network.

7.2.6.3 Application Layer

The application layer builds the ecosystem of an intent-driven campus network. The intent-driven campus network uses the SDN controller to provide standard northbound interfaces, so that service servers can program the northbound interfaces to the SDN controller and display application components on the controller.

The application layer contains standard applications provided by network vendors. Specifically, various network service control applications, network O&M applications, and network security applications constitute the SDN controller's standard functions. Additionally, to meet customized campus requirements, the SDN controller provides open third-party access functions. On the one hand, network vendors can cooperate with third-party partners to develop third-party applications and solutions (such as IoT solutions and business intelligence solutions). On the other

TABLE 7.5   Management Layer of the Intent-Driven Campus Network

Engine	Capability	Description
Intent engine	Simplicity	Enables administrators to use natural languages to manage networks. The intent engine automatically translates natural languages into executable languages and delivers configurations to network devices
Policy engine	Simplicity	Allows administrators to configure access rules and policies based on communications between people and people, between people and applications, as well as between applications and applications, instead of network elements such as IP network segments. The policy engine applies to people and applications, and is completely decoupled from the campus network, achieving maximum flexibility and ease of use
Security engine	Security	Frees administrators from worrying about network security. The security engine uses big data analysis and AI technologies to detect security threats and risks on the network through multidimensional risk evaluation. Once a security risk is detected, the security engine automatically makes contact with the intent engine to isolate or block the risk
Analytics engine	Intelligence	Enables administrators to proactively perform network O&M. The analytics engine uses technologies such as big data and AI to perform service correlation analysis, quickly locate and resolve network problems, and notify the administrator in advance before end users detect faults

hand, network vendors provide third-party customization functions, allowing customers to program custom applications.

7.2.6.4 WLAN for Intent-Driven Campus

During digital transformation, enterprises have long been aware of the convenience of wireless connections. Wireless access is the development trend of campus networks, and as such, WLAN will become the mainstream access mode. As compared to traditional campus networks, on an intent-driven campus network, a WLAN is still located at the access layer of the network layer. The difference is that APs are pooled at the network layer and managed as well as scheduled by the management layer in a unified manner, which greatly simplifies WLAN deployment and O&M. As such, all-scenario wireless solutions are implemented with the application layer. Table 7.6 compares WLAN deployment on intent-driven campuses and traditional campuses.

TABLE 7.6   WLAN Deployment on Intent-Driven and Traditional Campuses

Capability	Intent-Driven Campus Network	Traditional Campus Network
Ultra-broadband	The standard has evolved to Wi-Fi 6, with a maximum throughput of nearly 10 Gbps. With the 5G network's multiuser technology, orthogonal frequency division multiple access (OFDMA), the number of concurrent connections increases fourfold with latency reduction down to less than 20 ms	The AP has an internal IoT module, which integrates the WLAN with various IoTs to establish a fully connected network
Simplification	The SDN controller performs WLAN planning, design, deployment, and O&M, and automatically deploys service configurations. Accurate policy control is supported to ensure that policies and service experience are consistent when users move across the network	WLAN planning, design, deployment, and O&M require different platforms and tools Users need to log in to each WAC to configure services. The policy and service experience may be inconsistent when users move across the network
Intelligence	The WLAN status is visualized in real time, and the network trend is intelligently predicted based on historical data When the wireless environment changes, big data and AI can be used to quickly and accurately perform intelligent radio calibration Once a fault occurs, it can be quickly located while the administrator is notified. Some faults can even be automatically rectified without manual intervention	WLAN status cannot be detected in real time, and the network trend cannot be predicted After the radio environment changes, radio calibration cannot be performed promptly, and the calibration effect may not be satisfactory Fault location is difficult and requires highly skilled intervention, while troubleshooting is extremely time-consuming
Security	The device supports standard wireless security and encryption protocols, which authenticate wireless users as well as detect and prevent unauthorized wireless devices and attacks. In addition, big data technologies can be used to enable multidimensional analysis based on multiple scenarios on the entire network, detecting or even predicting threats and quickly handling them accordingly	Protection relies on the security functions of devices. For complex security attacks, extended periods of time are needed for analyzing logs and tracing sources. Manual policy configuration is required to handle detected threats, which is a time-consuming process

(Continued)

TABLE 7.6 (*Continued*) WLAN Deployment on Intent-Driven and Traditional Campuses

Capability	Intent-Driven Campus Network	Traditional Campus Network
Openness	The intent-driven campus network provides various APIs for the application layer, including authentication and authorization, location service, user profile, network O&M, and smart IoT services, enabling industry partners to quickly develop related applications	Solutions are independently developed by partners, Huawei, or both, leading to a long development period that cannot meet rapidly increasing industry application requirements

7.3 TYPICAL WLAN NETWORKING SOLUTIONS

7.3.1 Networking Solutions for Large Campuses

A large-sized campus includes the headquarters and branches of large- and medium-sized enterprises, universities, and airports. Large-scale campus networks require many APs. To ensure network O&M and security, the WAC+Fit AP architecture is used.

There are two WAC deployment solutions for large-scale campus networks: independent WAC and native WAC.

7.3.1.1 Independent WAC Solution

If a wired campus network has been deployed and a wireless network needs to be added, or the wireless network has a large scale, independent WAC deployment is recommended. For large-scale campus networks, the WAC is usually connected to the aggregation or core switch in bypass mode. To minimize changes to the existing wired network, tunnel forwarding is recommended to facilitate centralized management and control of WACs. To improve WAC reliability, VRRP hot standby mode is deployed in the independent WAC solution, as illustrated in Figure 7.22.

7.3.1.2 Native WAC Solution

A new campus network has numerous wired and wireless access devices. These devices are widely distributed, leading to complicated management and configuration on the access layer. Therefore, the native WAC solution is recommended for centralized management and configuration of wired and wireless access devices to reduce management costs. The Ethernet Network Processor (ENP) series native WACs serve as the core

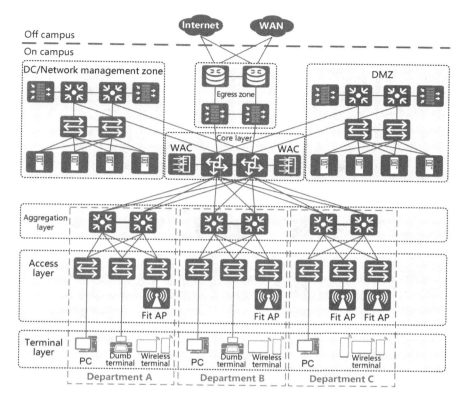

FIGURE 7.22   Independent WAC networking.

of this solution. In addition, the solution integrates the WAC function into switches. By installing special service boards on switches, they can be utilized by users to manage both wired and wireless access devices.

Native WACs can be used to provide network access for both wired and wireless users to implement unified management. The native WAC solution uses the reliability technologies (stacking and link aggregation) of switches to implement device-level and link-level redundancy, as illustrated in Figure 7.23.

7.3.2 Enterprise Branch Networking Solution

If both the headquarters and branches have WLANs deployed and the headquarters needs to manage the branch WLANs, enterprise branch WLAN networking can be used. Based on the network scale, enterprise networks can be classified into large-sized and small-sized enterprise branch networks.

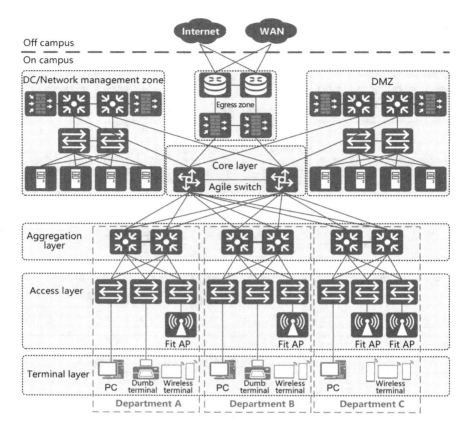

FIGURE 7.23   Native WAC networking.

7.3.2.1 Large-Sized Enterprise Branch WLAN Networking

On a large-sized enterprise branch network, a WAC is deployed to manage the wireless network of the branches. The headquarters' network management system configures and monitors the WLAN in a unified manner, as illustrated in Figure 7.24.

7.3.2.2 Small-Sized Branch WLAN Networking

In a small-sized enterprise branch network, the headquarters' WAC manages APs in both the headquarters and branches, as illustrated in Figure 7.25.

In this scenario, the headquarters and branches are connected through a WAN; the WAC and RADIUS server are deployed at the headquarters, and APs are deployed at the branches. In this case, the local forwarding

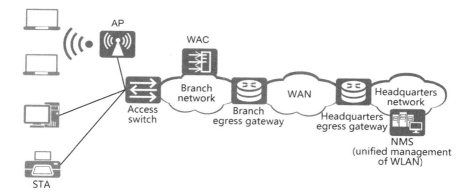

FIGURE 7.24   Branch networking of a large enterprise.

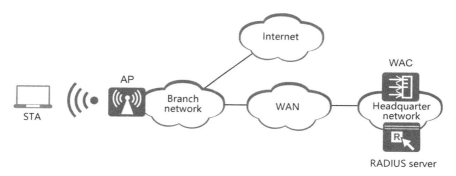

FIGURE 7.25   Branch networking of a small enterprise.

mode is used and the branch gateway assigns IP addresses to branch users for Internet access. If the headquarters and branches need to communicate with each other, an Internet Protocol Security (IPsec) virtual private network (VPN) tunnel is deployed between them.

7.3.3 Small- and Medium-Sized Enterprises and Small- and Micro-Branch Networking Solution

Small- and medium-sized enterprises usually have dozens to hundreds of employees. Small-sized enterprises, such as Small Office/Home Office (SOHO), have only a few to a dozen employees. Small- and micro-branches usually have several to dozens of employees. Enterprises such as supermarket chains, hotels, intermediaries, and regional sales offices might have widely distributed branches that serve a limited number of people within a specific area. Generally, a maximum of 50 APs are required for

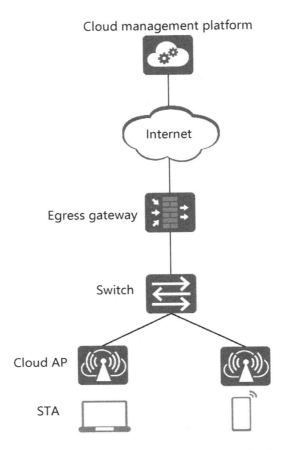

FIGURE 7.26 Small- and micro-branch networking using cloud management.

wireless network deployment, and this may lead to high overall costs if the WAC+Fit AP architecture is used. This is because WAC costs apply only to a small number of APs. Therefore, in this scenario, using the WAC-free cloud management architecture or leader AP architecture is recommended.

7.3.3.1 Small- and Micro-Branch Networking

Chain supermarkets and hotels have numerous branches located in various areas. The traditional network architecture has high deployment costs and difficult O&M, whereas these issues have been addressed in the cloud management architecture. Specifically, APs can be managed and maintained on the cloud management platform from any location, reducing network O&M costs. Refer to Figure 7.26.

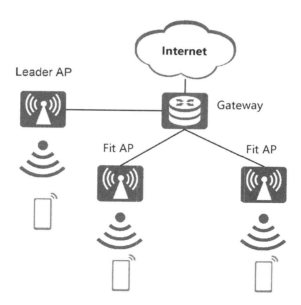

FIGURE 7.27   Small- and medium-sized enterprise networking using a leader AP.

7.3.3.2 Small- and Medium-Sized Enterprise Networking

Independent small- and medium-sized enterprises do not require central-ized management for multiple branches. Therefore, these enterprises can utilize the leader AP architecture to reduce network deployment costs and manage their own networks independently. The leader AP centrally man-ages fit APs locally, deploys wireless services, and maintains wireless net-works. Refer to Figure 7.27.

Enterprise WLAN Planning and Design

FOR LARGE-SCALE ENTERPRISE WIRELESS local area networks (WLANs), wireless network planning and design constitute a key determinant of network quality. Ideal network planning and design meet the requirements such as extensive coverage, conflict prevention, and large network capacity. This chapter describes WLAN planning and design methods based on the real-world network construction process.

8.1 NETWORK PLANNING AND DESIGN CONCEPT

When network planning and design are not conducted in a professional manner, certain issues may occur frequently after a WLAN project is delivered. Some of these issues commonly include:

1. Weak AP signals

    The actual environment is not considered during AP location planning, which may lead to signal coverage holes. For example, in a large office with a single AP coverage radius that exceeds 20 m, weak coverage will occur at the edge of the office, and the downlink signal strength detected by STAs will be lower than −75 dBm. This will lead to STA access failures or poor user experience.

2. Low Internet access rate on STAs

The WLAN uses the carrier sense multiple access with collision avoidance (CSMA/CA) mechanism. A larger number of concurrent users result in greater contention for channel resources and, consequently, a higher probability of conflict. For example, in an auditorium where seats are densely distributed, if a common dual-radio AP is used, excess users are carried on each frequency band, which leads to a high probability of packet conflicts among users and a low Internet access rate. In such scenarios, a triple-radio AP or a high-density directional antenna AP is required to improve network capacity.

3. Severe co-channel interference

Co-channel interference indicates that two APs that operate at the same frequency interfere with each other, resulting in poor network quality, slow network speed, or even service unavailability. Therefore, longer distances are more suitable for intra-frequency APs.

4. Poor experience in VIP areas

The services and experience of customers in VIP areas must be guaranteed, and WLAN planning needs to be taken into account on a case-by-case basis.

The preceding problems can be avoided through professional network planning and design. It is paramount to know in advance what types of wireless networks provide optimal user experience as well as how to measure and evaluate networks using quantitative indicators. In that regard, Huawei is proposing the high-quality WLAN concept and formulating corresponding network construction standards to guide the network planning and design process. See Figure 8.1. The core concept of a high-quality WLAN is "X Mbps anytime, anywhere."

1. X Mbps

Network planning must be geared toward the future in terms of meeting service requirements over the next 3–5 years. In enterprise office scenarios, with the popularization of Wi-Fi 6 standards, enterprise applications such as VR/AR, 4K/8K HD video, and all-wireless cloud desktop are being widely implemented at a rapid pace. These applications require existing network bandwidth to increase several times while still providing optimal experience to users and a fast

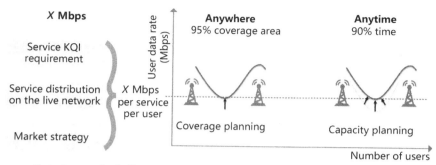

Note: key quality indicator (KQI)

FIGURE 8.1   Huawei high-quality WLAN.

network speed. Therefore, the core elements of X Mbps are the experience and endurance rates.

Experience rate: indicates the user-perceived rate in areas with good signal coverage and normal network load. This rate is essentially the same as that measured by the speed test software. In areas where signal coverage needs to be ensured, users can obtain higher experience rates. Whether VIP areas need to be planned depends on customers' actual network construction requirements.

Endurance rate: indicates the lowest rate during peak hours in all areas with continuous signal coverage (95% of the areas on the live network). The endurance rate determines the basic guarantee capability of the network.

In addition to the preceding two indicators, delay and coverage are also key performance indicators (KPIs) of the wireless network. Table 8.1 lists recommended KPI values.

TABLE 8.1   Recommended WLAN KPIs

KPI	Recommended Value
Experienced rate	>100 Mbps
Experienced rate in VIP areas	>300 Mbps
Endurance rate	>16 Mbps
Average downlink latency	High-priority applications <10 ms
	Low-priority applications <20 ms
Proportion of common coverage areas (uplink RSSI <−65 dBm)	<5%

2. Anytime

In all coverage areas, if the Internet access rate is not lower than 16 Mbps during 90% of the measurement period when network load is less than 80%, then the endurance rate is 16 Mbps; that is, a user can also obtain this rate in blocked areas or network edges.

In the central coverage area, if the Internet access rate is not lower than 100 Mbps in 90% of the measurement period when the network load is less than 50%, then the experienced rate in the good coverage area is 100 Mbps.

In a VIP area, when network load is less than 50%, a minimum experienced rate that is 300 Mbps during 90% of the measurement period is required. To achieve this rate, a more detailed network planning design must be implemented for the VIP area.

3. Anywhere

The network planning meets continuous coverage requirements. The signal coverage strength in 95% of the areas meets the experienced rate (100 Mbps) and endurance rate (16 Mbps) requirements.

The following sections describe how to construct high-quality wireless networks with the preceding characteristics.

8.2 NETWORK PLANNING PROCESS

Figure 8.2 illustrates the WLAN planning process.

1. Collect basic project information and requirements from customers.

2. Perform site surveys and collect information about interference sources as well as network planning and design obstacles.

3. Perform network planning and design.

Based on site survey results and customer requirements, engineers determine the WLAN coverage mode (indoor coverage, outdoor coverage, high-density coverage, or backhaul), and then design the network planning solution from network coverage, network capacity, and AP deployment.

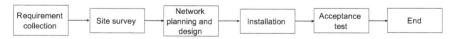

FIGURE 8.2   Network planning process.

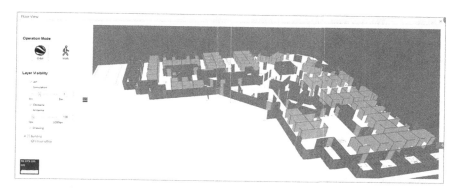

FIGURE 8.3   3D simulation of Huawei's cloud-based WLAN Planner.

Professional software tools, such as Huawei cloud-based WLAN Planner, can be used to design network planning solutions and simulate network coverage. After importing the drawing and adding obstacle information, the software automatically deploys APs and conducts 3D simulation of the actual space. The software can also simulate the coverage effect of each point on a specified route. Through this process, users can intuitively determine whether the network plan meets coverage requirements. See Figure 8.3.

4. Install the equipment according to the network planning and design scheme.

5. Perform an acceptance test after the construction is complete. The specific content of the test includes whether network coverage and service usage requirements have been met. Network test tools can be used to perform the acceptance test. For example, the CloudCampus app can test the coverage effect of each area and check whether common services are running smoothly. After the acceptance is passed, the entire network planning process ends.

8.3 REQUIREMENT COLLECTION AND ONSITE SURVEY

8.3.1 Requirement Collection

The first step of WLAN planning is to collect relevant requirements. Communicate with customers to clarify their requirements and specify network construction objectives. For details about the type of information that needs to be collected, see Table 8.2.

TABLE 8.2 WLAN Planning Requirements

Requirement	Description
Regulatory restrictions	Check the equivalent isotropically radiated power (EIRP) restriction and available channels. For example, under China's country code, channels 1–13 (EIRP 27) support the 2.4 GHz frequency band, and channels 36–64 (EIRP 23) and channels 149–165 (EIRP 27) support the 5 GHz frequency band
Drawing information	Check the integrity of the drawings that contain the scale information. Usually, the construction CAD drawings can be obtained from a customer's construction management office
Coverage area	Confirm key and common coverage areas with the customer. Key coverage areas include office areas and conference rooms, and common coverage areas include staircases and bathrooms
Signal strength	Confirm the customer's signal strength requirements within the respective coverage areas. For example, −65 to −40 dBm in key areas, and −75 dBm or higher in common areas
Number of users	Determine the number of users and calculate the total number of associated STAs in a coverage area. In wireless office scenarios, assuming that each person uses one mobile phone and one laptop, the total number of associated STAs is calculated as follows: Number of associated STAs = Number of users × 2
STA type	Confirm STA type and quantity. Common STAs include mobile phones, tablets, and laptops, while special terminals include barcode scanners and cash registers. Calculate the proportion of STAs of different multiple-input multiple-output (MIMO) types to estimate the required AP performance (optional, based on customers' technical capabilities)
Bandwidth	Confirm the major service types planned by the customer, and the bandwidth requirement of each user
Coverage mode	Check whether the customer requires indoor settled, agile distributed, or outdoor APs for coverage
Power supply mode	Check whether the customer has specific requirements for the power supply mode as well as whether power supply areas and on-site facilities are available
Switch location	Determine the switch positions on the upstream wired-side from WLAN devices

8.3.2 Site Survey

WLAN planning is based on-site surveys, which involve obtaining information such as STA type, floor height, building material and signal attenuation, interference source, new obstacles, and weak-current well location. This information helps determine AP model, AP installation mode, AP location, power supply, and cabling design. Generally, site surveys are completed by using tools. Table 8.3 describes common tools, whereas Figure 8.4 shows them.

TABLE 8.3   Common Site Survey Tools

Type	Name	Description
Software	On-Pad Surveyor (Android)	• Mark the obstacle position and type, test the signal attenuation, and record the result • Mark the location and type of the interference source, detect the AP interference source, and record the channel and transmit power • Add photo and text annotations to drawings • Modify the scale and floor attributes • Import project data between the On-Pad Surveyor and WLAN Planner
	CloudCampus app	CloudCampus app is a convenient Wi-Fi signal detection tool that enables users to detect network speed and security. It provides multiple functions such as AP locating, STA checks, and interference checks. In addition, the app's survey module supports multiple recording modes for information in text, voice, and image form
	inSSIDer	inSSIDer is a third party wireless network signal scanning tool. Users can use inSSIDer to quickly locate hotspots in their vicinity and collect information such as MAC addresses, network names, encryption modes, and wireless transmission rates
	WLAN Planner	WLAN Planner is a professional wireless network planning tool developed by Huawei. It is used to survey the on-site environment, plan AP installation locations, simulate network signals, and generate reports. This tool helps users easily plan wireless networks and deploy APs, improving work efficiency
	AutoNavi map or Google Earth	(For outdoor scenarios) Mark the longitude and latitude of an AP, check obstacles, and confirm the project environment
Hardware	Indoor rangefinder	(For indoor scenarios) Measure the AP installation height, distance between the AP and obstacles, as well as the venue's length, width, and height
	Camera	Record site environment information, such as the AP installation environment and obstacle information in WDS networking scenarios. In indoor scenarios, use the On-Pad Surveyor to complete the survey. In outdoor scenarios, it is recommended that you carry a camera

(Continued)

TABLE 8.3 (*Continued*)   Common Site Survey Tools

Type	Name	Description
	Test AP (including the auxiliary power supply and tripod)	Use the On-Pad Surveyor to test the signal attenuation of obstacles in indoor scenarios When using the test AP (FAT mode), you need to take a power supply and tripod (upwards of 2 m to simulate the ceiling-mounted installation scenario)
Others	Building drawing	It is recommended that you print the project drawings in advance to facilitate site survey

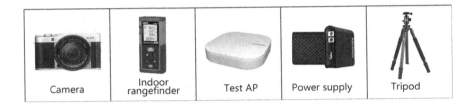

| Camera | Indoor rangefinder | Test AP | Power supply | Tripod |

FIGURE 8.4   Common site survey tools.

8.3.2.1 Tools

8.3.2.2 Site Survey Items

In complex scenarios, a site survey may take a long period of time. To avoid omitting details, refer to the site survey information collection table while performing the site survey. It is recommended that the site survey information be recorded in detail at the corresponding position on the drawing.

Table 8.4 uses the enterprise office scenario as an example to describe site survey information to be collected in indoor scenarios.

In a wireless network environment, obstacles cause user experience to deteriorate. During site surveys, you need to focus on and know how to test signal attenuation caused by obstacles. The following section describes how to test signal attenuation using tools.

8.3.2.3 Signal Attenuation Caused by Obstacles

Step 1: Select an obstacle to be tested.

Generally, the selected object is a typical indoor obstacle or an obstacle made of unknown material (such as a ceiling or decorated wall).

Step 2: Place and enable the test AP (FAT mode).

TABLE 8.4   Site Survey Information

Item	Example	Note
Floor height	The height of a common indoor floor is 3 m	Obtain the ceiling and atrium height of a lobby or lecture hall Use an indoor rangefinder to measure the distance
Building materials and signal attenuation	240 mm brick wall (2.4 GHz: 15 dB attenuation, 5 GHz: 25 dB attenuation), 12 mm thick colored glass (2.4 GHz: 8 dB attenuation, 5 GHz: 10 dB attenuation)	Obtain the thickness and signal attenuation value of the building materials on site. If possible, test the signal attenuation value on site If the test cannot be performed, refer to the attenuation value in Table 8.6
Interference sources	Wi-Fi interference is detected, and the interference source information has been marked on the drawing	Check whether there is interference caused by, for example, mobile hotspots, other vendors' Wi-Fi devices, and non-Wi-Fi devices (such as Bluetooth devices and microwave ovens). CloudCampus app can be used to record interference source information
New obstacles	New obstacles on the site have been marked on the drawing	Check whether the site is consistent with the building drawing. If not, mark the key areas and take photos to record the indoor landscape
Site photos	Site photos are required	Take photos of the site to record environment information for further site survey
AP type	Indoor settled APs are used	Select indoor settled APs, agile distributed APs, outdoor APs, or high-density APs based on specific scenarios
Installation mode and location	APs are ceiling or wall-mounted	Check whether APs can be mounted on the ceiling. If ceiling mounting is unavailable, mount it on a wall or panel
Weak-current well location	The weak-current well locations have been marked on the drawings	Mark the locations of weak-current well where switches are to be deployed on the drawings
Power supply cabling	Power supply cabling has been marked on the drawing	Mark the power over Ethernet (PoE) cabling on the drawing. A PoE network cable length that is less than or equal to 80 m is recommended
Special requirements	The roaming packet loss rate is less than 1%, and the delay is less than 20 ms	Special user requirements, such as roaming packet loss rate and delay
Others	None	Others

Note: For areas with similar spatial structures, you can select only one typical test area.

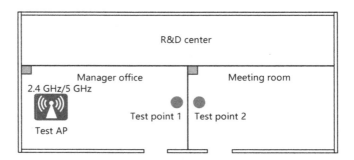

FIGURE 8.5   Obstacle attenuation test.

Ensure that there is no object between the test AP and obstacle. The distance between the test AP and obstacle is 4–5 m. Do not place the test AP close to the obstacle because the signal strength near a signal source fluctuates significantly, which affects test accuracy.

Step 3: Use a signal scanning tool (CloudCampus app on a mobile phone or inSSIDer on a laptop) to test the signal strength on both sides of the obstacle. The signal attenuation value is the difference between the signal strength values. You are advised to measure several groups of data to reduce the possibility of errors occurring.

As illustrated in Figure 8.5, after the 2.4 and 5 GHz radio parameters are configured for the test AP, the signal strengths for 2.4 and 5 GHz measured at test point 1 are −50 dBm, while those for 2.4 and 5 GHz measured at test point 2 are −60 and −65 dBm, respectively. Therefore, the signal attenuation value of the obstacle for 2.4 GHz is 10 dB, and that for 5 GHz is 15 dB.

Step 4: Log the obtained signal attenuation value into the WLAN Planner, as illustrated in Figure 8.6.

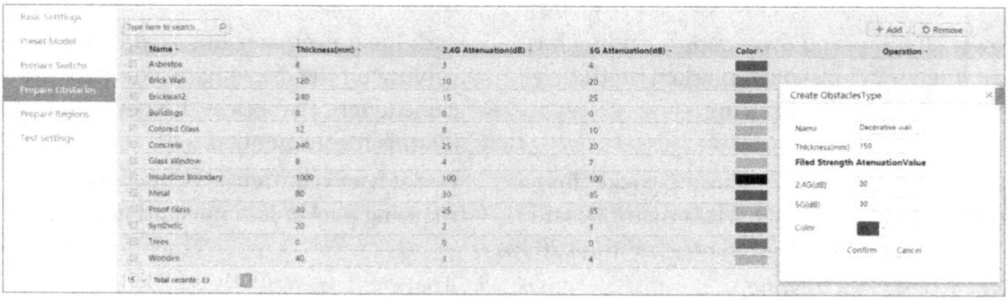

FIGURE 8.6   Customizing an obstacle on the WLAN Planner.

8.4 NETWORK COVERAGE DESIGN

To improve user experience, you need to consider the signal strength requirements in the wireless network coverage area during network planning. Table 8.5 lists signal strength requirements for typical scenarios.

High-quality networks pose higher requirements on signal strength in common coverage areas. Specifically, signal strength must be greater than −65 dBm; therefore, APs with smart antennas are recommended. Beamforming of smart antennas ensures that signal strength in the coverage area meets requirements and avoids co-channel interference between APs.

In an open area without obstacles, the coverage radius of a single AP can reach 15 m. If an obstacle exists, the coverage radius decreases based on the obstacle's signal attenuation. In this case, the actual signal strength may be obtained according to the following formula for calculating signal strength.

$$\text{Signal strength} = \text{AP transmit power} + \text{Antenna gain}$$

$$- \text{Transmission distance attenuation}$$

$$- \text{Signal attenuation caused by obstacles}$$

When the signal transmission distance is 15 m without considering factors such as interference and obstacles, the signal strength at 5 GHz is as follows:

$$\text{AP transmit power} \ (20 \ \text{dBm}) + \text{Antenna gain} \ (4 \ \text{dBi})$$

$$- \text{Transmission distance attenuation} \ (88.3 \ \text{dB})$$

$$- \text{Signal attenuation caused by obstacles} \ (0 \ \text{dB}) = -64.3 \ \text{dBm}$$

TABLE 8.5   Mapping between Coverage Areas and Signal Strengths

Coverage	Signal Strength Requirements	Common Project Areas
Major coverage areas	−40 to −65 dBm	Dormitory, library, classroom, hotel room, lobby, meeting room, office, and exhibition hall
Common coverage areas	>−75 dBm	Corridor, kitchen, storeroom, and dressing room
Special coverage areas	Based on site conditions	Area that is specified or does not allow coverage or installation

TABLE 8.6   Mapping between Signal Transmission Distances and Attenuation Values

Distance (m)	2.4 GHz Signal Attenuation (dB)	5 GHz Signal Attenuation (dB)
1	46	53
2	53.5	62
5	63.5	74
10	71	83
15	75.4	88.3
20	78.5	92
40	86	101
80	93.6	110.1

TABLE 8.7   Signal Attenuation Values of Common Obstacles

Typical Obstacle (Material)	Thickness (mm)	2.4 GHz Signal Attenuation (dB)	5 GHz Signal Attenuation (dB)
Synthetic material	20	2	3
Asbestos	8	3	4
Wood door	40	3	4
Glass window	8	4	7
Heavy colored glass	12	8	10
Brick wall	120	10	20
Brick wall	240	15	25
Armored glass	120	25	35
Concrete	240	25	30
Metal	80	30	35

Table 8.6 lists the mapping between signal transmission distances and attenuation values in indoor scenarios. Table 8.7 lists the signal attenuation values of common obstacles.

The data in Table 8.7 has been preset in the WLAN Planner, enabling users to plan APs and simulate heat maps. You can also test the signal attenuation of unknown obstacles by using the preceding method and recording test results in the network planning tool.

8.5 NETWORK CAPACITY DESIGN

Wireless network capacity is based on actual service requirements. That is, the number of APs required is based on AP performance, bandwidth requirements (endurance rate), number of users, and wireless environment. In campuses, offices, and education scenarios, more and more services are being carried on wireless networks, requiring service bandwidth to increase year by year. Therefore, when designing networks in these scenarios, service requirements for the next 3–5 years must be also taken into account. In addition, in scenarios such as hotels, shopping malls, supermarkets, and outdoor coverage areas, the network capacity can be designed based on actual service requirements, with both service requirements and costs considered.

8.5.1 Single-AP Performance

Table 8.8 lists the recommended typical number of concurrent users for a single AP that works in 802.11ac 20 MHz mode with different single-user bandwidth requirements. In this table, data at 2.4 and 5 GHz is based on the 802.11n standard and 802.11ac standard, respectively.

Table 8.9 lists the typical number of concurrent users recommended for a single AP when the AP works in 802.11ax 40 MHz mode and different single-user bandwidth requirements are present.

8.5.2 WLAN STA Concurrency Rate in Typical Scenarios

WLAN usage varies between different scenarios. In wireless office scenarios, for example, where each user possesses one laptop and one mobile phone, the STA concurrency rate is considered 40% if the service type is not specified. The WLAN STA concurrency rate in common scenarios is illustrated in Table 8.10.

8.5.3 Single-User Bandwidth Requirement Estimation

The bandwidth required by a user depends on the service type. The typical bandwidth requirements of common network applications are illustrated in Table 8.11.

As a user employs multiple services at a time, to accurately calculate the bandwidth required by a user, the usage proportion of each service in a period of time must be obtained. Then, the bandwidth required by each service is multiplied by the usage proportion, and the results added. Although this equation provides an accurate calculation, it can be too complex. In scenarios where sufficient bandwidth is present, a service that requires the maximum bandwidth is selected for calculation, and the calculated bandwidth is used as the bandwidth requirement of a single user. For example, in office scenarios where 16,000 kbps is required to complete a mail or file transfer, that specified amount (16,000 kbps) may be used as the calculated bandwidth required by a single user.

TABLE 8.8 Typical Number of Concurrent Users Supported by an AP with Different Single-User Bandwidth Requirements in 802.11ac 20 MHz Mode

Single User Bandwidth (Mbps)	1 × 1 SISO (STA Performance)			2 × 2 MIMO (STA Performance)		
	Number of Concurrent Users on 5 GHz	Number of Concurrent Users on 2.4 GHz	Number of Concurrent Users on Both 2.4 and 5 GHz	Number of Concurrent Users on 5 GHz	Number of Concurrent Users on 2.4 GHz	Number of Concurrent Users on Both 2.4 and 5 GHz
16	2	2	4	5	4	9
8	5	4	9	9	7	16
6	6	5	11	11	8	19
4	8	7	15	14	10	24
2	12	10	22	20	16	36
1	20	15	35	30	25	55
0.5	30	25	55	35	30	65

Note: The data in Table 8.8 is obtained using STAs that support the 802.11ac standard, while the 2.4 GHz frequency band is configured to the 802.11n 20 MHz mode and the 5 GHz frequency band is configured to the 802.11ac 20 MHz mode in the dual-band mode.

TABLE 8.9 Typical Number of Concurrent Users Supported by an AP with Different Single-User Bandwidth Requirements in 802.11ax 40 MHz Mode

Single User Bandwidth (Mbps)	1 × 1 SISO (STA Performance)			2 × 2 MIMO (STA Performance)		
	Number of Concurrent Users on 5 GHz	Number of Concurrent Users on 2.4 GHz	Number of Concurrent Users on Both 2.4 and 5 GHz	Number of Concurrent Users on 5 GHz	Number of Concurrent Users on 2.4 GHz	Number of Concurrent Users on Both 2.4 and 5 GHz
16	8	3	11	10	6	16
8	16	6	22	20	12	32
6	18	8	26	22	14	36
4	24	11	35	30	15	45
2	33	15	48	45	30	75

Note: The test STA in Table 8.9 must support the 802.11ax standard.

TABLE 8.10   Reference Values of the WLAN STA Concurrency Rate

Scenario	Scenario Description	Concurrency Level of WLAN Users	Reference Concurrency Rate
Office	There is wired broadband access, and WLAN access is mainly for mobile phones	Medium	25%–40%
Office	There is no wired broadband access, and each person has one laptop and one mobile phone for WLAN access	High	>40%
Multimedia classroom	WLAN access is mainly for tablets	High	100%
Student dormitory	There are a small number of wired broadband connections. The STA penetration rate is 100%	High	>50%
Exhibition hall	There are a few wired interfaces. WLAN access is mainly for mobile phones but also for a small number of laptops and tablets	Medium	25%–40%
Hotel	There is wired broadband access. WLAN access is mainly for laptops but also for a small number of mobile phones	Medium	25%–40%
Street and entertainment venue	WLAN access is mainly for mobile phones and tablets	Low	20%–30%
Stadium	WLAN access is mainly for mobile phones and tablets	Low	20%–30%
Traffic hub	WLAN access is primarily for mobile phones and tablets but also for a few laptops	Low	10%–20%
Campus	WLAN access is primarily for mobile phones and tablets but also for a few laptops	Low	20%–30%

Note: As service types vary in different scenarios, during capacity design, the service type, STA concurrency rate, and online user activity level must be considered to determine a reliable user concurrency rate.

However, considering the rapid growth of high-bandwidth services, such as live conferences, it is recommended that some bandwidth be left in reserve during bandwidth planning to ensure efficient service evolution in the future.

8.5.4 AP Quantity Estimation

To estimate the number of APs needed to establish an optimized network, coverage requirements and capacity requirements need to be considered.

TABLE 8.11   Typical Bandwidths for Different Service Types

Service Type	Typical Bandwidth (kbps)
Web page browsing	4000
Video (480p)	3200
Video (720p)	6400
Video (1080p)	12,000
Video (4K)	22,500
VoIP (audio)	128
VoIP (video)	6400
Email	16,000
File transfer	16,000
Social network website	1200
Instant messaging	256
Mobile game	1000
VR video	40,000

Note: VoIP, short for Voice over Internet Protocol, is an IP-based voice
transmission protocol that can also transmit fax, video, and data.
The above data is provided by Huawei iLab and displays the bandwidth
of each service with good WLAN experience.

- To meet coverage requirements, calculate the distance between APs that meet different edge signal strength requirements based on Section 8.3, and then calculate the number of required APs based on the area's total space.

- To meet capacity requirements, calculate the number of APs using the below formula. In this formula, the performance of a single AP refers to the number of concurrent users in relation to the bandwidth of a single user. The data in Table 8.11 can be used to estimate the bandwidth required by a single user based on the actual service scenario. Then, using the data in Table 8.8, or Table 8.9, the number of concurrent AP users corresponding to the bandwidth can be found.

$$\text{Number of required APs} = \frac{\text{Number of associated STAs} \times \text{Concurrency rate}}{\text{Performance of a single AP}}$$

Calculate the numbers of required APs based on the signal coverage and capacity requirements, and choose the larger value as the number of APs estimated in the solution. As most enterprise WLAN scenarios are capacity-limited, the number of required APs can be estimated based on capacity requirements. By contrast, in scenarios with a small number of users and small bandwidth, such as factory production lines, the number of APs must be estimated based on coverage requirements.

8.6 AP DEPLOYMENT DESIGN

8.6.1 AP Deployment Scenarios

When APs are deployed in indoor and outdoor scenarios, site conditions need to be considered. Fortunately, in most indoor network planning scenarios, the general principles for AP deployment remain the same.

- Reduce the number of times that signals pass through obstacles, such as walls and ceilings, while also ensuring that signals can vertically pass through obstacles.

- Ensure that the front of the AP faces the target coverage area. If only one AP is deployed indoors, deploy it in the center of the indoor area. By contrast, if two APs are deployed, they can be placed on two diagonals. If the indoor area's floor plan changes, AP deployment direction can be easily adjusted.

- Deploy APs far away from interference sources and electronic devices. Devices such as microwave ovens, wireless cameras, Wi-Fi enabled phones, or other electronic equipment should not be deployed in the coverage area.

The following describes typical AP deployment scenarios and design solutions:

1. Common indoor scenario

   Indoor scenarios, such as classrooms, conference rooms, and offices, are characterized by their simple building structure and limited obstacles. In most cases, common APs with omnidirectional antennas are deployed.

   There are three scenarios based on room size, as illustrated in Figure 8.7.

   - If the area of a single room is less than 50 m², only one AP needs to be installed.

   - If the area of a single room ranges from 50 to 100 m², two APs need to be installed.

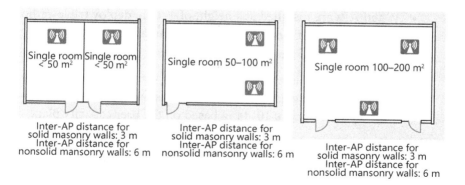

FIGURE 8.7   AP deployment in common indoor scenarios.

- If the indoor room area is greater than 100 m², calculate the number of required APs based on the actual situation and deploy the APs in an equilateral triangle.

2. Dense-room scenario

In dense-room scenarios, including hotel rooms, dormitories, and hospital wards, the room area is characterized as small, with users densely distributed. In this case, the agile distributed solution is used, as illustrated in Figure 8.8.

3. High-density scenario

In high-density scenarios, such as stadiums, large conference rooms, and ticket halls, users are densely populated in large areas.

In scenarios like this, deployment rules are based on the AP installation height.

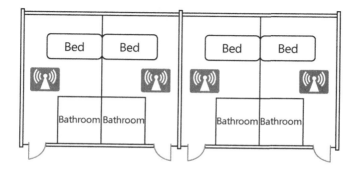

FIGURE 8.8   AP deployment solution in dense-room scenarios.

- When APs are installed at a height of 4–6 m, deploy triple-radio settled APs with omnidirectional antennas with a separating distance of 10–15 m.

- When APs are installed at a height of 6–10 m, deploy APs with built-in small-angle directional antennas (e.g., 30° × 30°), separated by a distance of 10–14 m based on the user capacity plan.

- When APs are installed at a height of 10–25 m, deploy APs with external small-angle directional antennas (e.g., 15° × 15°), separated by a distance of 8–16 m based on the user capacity plan.

- When APs are installed at a height greater than 25 m, deploy APs with external small-angle directional antennas for coverage. The distance between APs is calculated based on the antenna type and user capacity.

4. Indoor public areas

   In indoor public area scenarios, including indoor corridors, coverage is the primary requirement. Therefore, it is recommended that settled APs with omnidirectional antennas be installed on the indoor area's ceilings. It is also recommended that APs provide linear coverage with an equal spacing of 20 m, as illustrated in Figure 8.9.

5. Outdoor scenarios

   Outdoor scenarios are areas with open spaces and a few obstacles, such as parks, public squares, and streets.

   - For parks, APs with omnidirectional antennas and spacing of 50–60 m are used for coverage, as illustrated in Figure 8.10 (left).

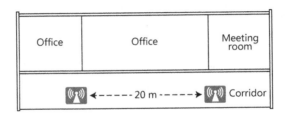

FIGURE 8.9   AP deployment in indoor public areas.

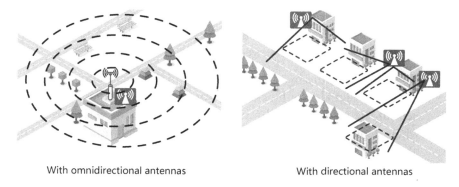

With omnidirectional antennas   With directional antennas

FIGURE 8.10   AP deployment in parks and squares.

- For public squares, APs with large-angle directional antennas are used for coverage. APs are mounted on the external walls of buildings at a height of 4–6 m, and the distance between APs is 30–40 m, as illustrated in Figure 8.10 (right).

- For roads, as illustrated in Figure 8.11 APs with directional antennas are used for coverage. The distance between APs is 120–150 m.

In addition to the preceding AP deployment design principles, the following installation mode requirements must be satisfied:

- Aesthetic considerations: In common indoor scenarios without high aesthetic requirements, APs can be directly installed. However, in high-end office areas, APs are installed inside the ceiling (usually nonmetal) with a camouflage cover.

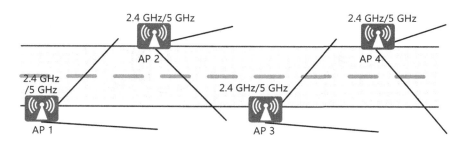

FIGURE 8.11   AP deployment along roads.

- An overlapping roaming area: In areas where roaming is required, the coverage areas of neighboring APs must overlap by 10%–15% to ensure smooth transfer between APs.

- No signal blocking by obstacles: When an AP is close to a column and radio signals are blocked, a large coverage shadow is formed behind the column. To eliminate coverage holes or weak coverage when deploying APs, avoid deployment near columns and metal objects that have a strong reflection effect on wireless signals. What's more, do not place APs or antennas behind metal ceilings.

8.6.2 Channel Planning

To reduce interference between APs, pay attention to channel planning in scenarios where rooms are densely distributed. As illustrated in Figure 8.12, APs on the same channel must be separated by the longest possible distance.

To improve AP deployment and network planning efficiency, the WLAN Planner can be used to automatically plan AP channels. In addition, the WAC can be used to support automatic radio calibration, and it can specify the channel and power of APs based on detected interference and neighboring AP information, ultimately simplifying network configuration operations.

In building scenarios, inter-floor interference needs to be reduced to avoid channel conflict and unify channel planning, as illustrated in Figure 8.13.

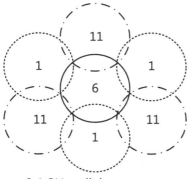

2.4 GHz cellular coverage

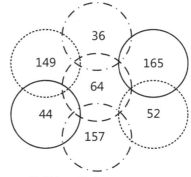

5 GHz cellular coverage

FIGURE 8.12 Typical channel planning.

Floor	Channel Plan		
F7	1	6	11
F6	11	1	6
F5	6	11	1
F4	1	6	11
F3	11	1	6
F2	6	11	1
F1	1	6	11

FIGURE 8.13   Channel planning for a multistory building.

8.7 POWER SUPPLY AND CABLING DESIGN

8.7.1 Power Supply

APs support power supply using PoE or direct current (DC) power adapters.

8.7.1.1 PoE Power Supply

PoE power supply is recommended in most application scenarios, such as enterprise offices, classrooms, dormitories, and stadiums.

PoE allows network cables to supply power to APs in addition to data transmission. It features easy deployment, quick construction, and stable and secure power supply.

PoE power supplies are usually provided by widely available PoE switches (see Figure 8.14). PoE switches, usually deployed in the central equipment room or intermediate node cabinet, include different models that support 8-, 16-, 24-, 32-, and 64-port configurations, meeting the requirements for connecting different numbers of APs.

After considering cable loss, the 802.3 standard family of the Institute of Electrical and Electronics Engineers (IEEE) has adopted network cable specifications required by ANSI/TIA/EIA-568-B, which defines cable types such as CAT5e, CAT6, and CAT6A to meet different transmission bandwidth and metal wire diameter requirements for a transmission distance of 100 m, as illustrated in Table 8.12.

FIGURE 8.14   PoE switch.

TABLE 8.12   Network Cable Types and Parameters

Cable Type	Transmission Bandwidth	Ethernet Transmission Rate	Metal Wire Diameter
CAT5e	100 MHz	100 Mbps, 1 , 2.5 Gbps	22–24 AWG
CAT6	250 MHz	5 Gbps	22–24 AWG
CAT6A	500 MHz	10 Gbps	22–24 AWG

Note: The IEEE 802.3bz standard specifies that the Alien Limited Signal-to-Noise Ratio (ALSNR) of the external crosstalk between network cables must be greater than 0. In this case, when the rate is 5 or 10 Gbps, you need to use shielded network cables with corresponding specifications to prevent issues such as continuous packet loss. American Wire Gauge (AWG) is a code defined by the Electronic Industry Association (EIA) for the diameter of metal conductors of network cables. A smaller value indicates a larger conductor diameter.

The IEEE 802.3af, 802.3at, and 802.3bt standards are for PoE power supply and define the DC resistance of network cables within 100 m according to the ISO/IEC 11801 standard, as illustrated in Table 8.13. The requirements of these standards are also met by the CAT5e, CAT6, and CAT6A cables defined by ANSI/TIA/EIA-568-B. That is, when network cables are used as a power supply medium, the DC resistance generates voltage drop, consuming the power of the power supply end. In this case, a network cable with a smaller DC resistance consumes less power of the system.

TABLE 8.13   Requirements of PoE Power Supply Standards for DC Resistance and Output Power of Network Cables

Standard	100 m Network Cable DC Resistance (Ω)	Output Power (W)
802.3af	20	≤15.4
802.3at	12.5	≤30
802.3bt	12.5	≤90

When selecting a PoE switch, it is important to check whether the transmission rate and power supply standard of the PoE port match those of the AP. For example, if an AP that supports Wi-Fi 6 has a maximum transmission rate of 6 Gbps and maximum power consumption of 30 W, the PoE port of the PoE switch must support a transmission rate of 10 Gbps to maximize the AP performance. What's more, the power supply standard must support 802.3at or 802.3bt, ensuring the power supply to the AP.

About 50%–60% of the total workload in network deployment is spent routing network cables and involves engineering activities, such as wall penetration, cabling in protective pipes, and buried cable installation. Therefore, high-specification network cables are recommended to accommodate future needs. In addition, due to signal crosstalk and twist in network cables, jumpers may be used in cable connections. As such, it is recommended that the network cable length be less than or equal to 80 m.

In actual networking, PoE power modules (illustrated in Figure 8.15) can be used to supply power to outdoor APs and some indoor APs that cannot be powered by PoE switches. Note that the PoE power module does not function as a network node and only functions as a network relay. The total length of network cables at both ends cannot exceed the network node requirement.

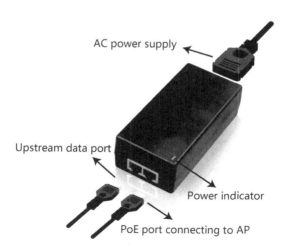

FIGURE 8.15   PoE power module.

In outdoor application scenarios, power supply and data transmission of an AP are separate. The PoE power supply module is connected to the nearest power grid (AC power). However, the surrounding network cables may not be sufficiently long for APs. In this case, optical fibers are used for data transmission, as optical fiber transmission can effectively increase the transmission distance between network nodes. For example, when a multimode optical module and a multimode optical fiber are used in conjunction, the transmission distance can reach up to 550 m. And when single-mode optical modules of different power levels are used with single-mode optical fibers, the transmission distance can reach 2, 10, or even 80 km.

8.7.1.2 DC Power Supply Using Adapters

If PoE power supply is unavailable at an indoor location, DC power supply (as illustrated in Figure 8.16) can be used. For indoor locations, you can choose a proper power adapter to supply power to the AP. However, the AP still needs to use a network cable as the media for connecting to the upstream network.

8.7.2 Cabling Design

During the construction, select an appropriate AP position based on the network plan, determine the upstream network node according to the preceding principles, and finally, deploy network cables. The following needs to be considered when deploying network cables:

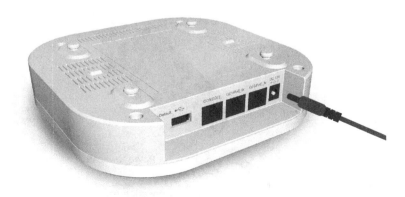

FIGURE 8.16   DC power supply.

- During AP deployment, reserve around 5 m cable length surplus for adjusting AP installation positions due to interference or poor signal coverage in the future.

- The network cable must be far away from strong electric and magnetic interference sources.

- Confirm with the customer about the network cable deployment scheme in advance to avoid property and aesthetic violations.

CHAPTER 9

Scenario-Based Enterprise WLAN Design

T HE PREVIOUS CHAPTER DESCRIBES general wireless local area network (WLAN) planning and design methods. This chapter analyzes typical scenarios and provides WLAN planning and design methods, enabling network administrators to better understand and use WLAN planning and design in actual projects.

9.1 ENTERPRISE OFFICES

The enterprise office scenario refers to the office area of an enterprise and includes areas such as meeting rooms and manager offices. It is one of the most important application scenarios of enterprise WLAN, and it features high user density, high requirement for network capacity, and high sensitivity to network quality.

9.1.1 Business Characteristics

1. Space characteristics

Figure 9.1 shows the enterprise office scenario, which has the following key features:

- Space: The height does not exceed 4 m, and the area varies from several square meters to thousands of square meters.

FIGURE 9.1   Enterprise offices.

- Blocking: Obstacles, such as cubicles and support pillars, are common in small and large offices. These obstacles can potentially block signals.

- Interference: The external interference source is rare in independent office areas. However, if several companies lease offices on the same floor, the WLANs of multiple companies may be severely interfered.

2. Service characteristics

In enterprise offices, the main STA types are laptops (including wireless network interface cards) and wireless office devices (such as printers and electronic whiteboards). Applications are classified into the following types:

- Office personal applications: office software, instant messaging software, email, file transfer, desktop sharing, and desktop cloud. Employees use these applications on laptops, office PCs, and tablets to access the enterprise intranet.

- Nonoffice personal applications: videos, games, and social software. Employees use these applications on their mobile phones to access the Internet.

- Enterprise IoT applications: asset management and energy efficiency control (air conditioners and lights).

9.1.2 Best Practices of Network Design

Based on the service characteristics of enterprise offices, the best practices of network design are as follows. Table 9.1 lists the involved features.

TABLE 9.1   Features in Enterprise Offices

Type	Feature Name	Feature Description
High-quality wireless network	AP performance	The AP complies with Wi-Fi 6 and supports 12 spatial streams (2.4 GHz 4×4 MIMO and 5 GHz 8×8 MIMO). What's more, the eight spatial streams on the 5 GHz frequency band can maximize the performance of Wi-Fi 6
	Multiuser multiple-input multiple-output (MU-MIMO)	As the number of spatial streams (one or two) of an STA is smaller than that of an AP, a single STA cannot fully utilize the performance of an AP, and MU-MIMO technology is introduced to allow multiple STAs to access one AP for data transmission. After MU-MIMO enhancement is enabled, the capacity can be doubled at least
	Load balancing	Enterprise offices are a typical indoor high-density scenario. And when a large number of STAs access the network, load balancing needs to be performed between APs and between 2.4 and 5 GHz frequency bands of the same AP
	Smart antennas	Huawei smart antenna technology increases the AP coverage by approximately 15% and reduces interference to other APs and STAs through beamforming
	Smart roaming	Employees and guests can roam to different areas. The pairwise master key (PMK) cache, 802.11r fast roaming, and network-initiated proactive roaming guidance are supported. The proactive roaming guidance technology enables the network side to monitor the STA link status in real time. And when the link quality deteriorates, the network side proactively directs STAs to a new AP to prevent STAs from being connected to the original AP with poor signal quality. What's more, the success rate of Huawei smart roaming technology is higher than 80%. This technology can identify sticky connection of STAs within 5 seconds and guide terminal roaming when needed
	Service-based quality of service (QoS) control	Enterprise office applications are classified into office and personal applications. That is, different applications have different QoS requirements on the network. To ensure user experience of key services, such as Voice over Internet Protocol and electronic whiteboard, QoS control must be performed based on service types

(Continued)

TABLE 9.1 (*Continued*)   Features in Enterprise Offices

Type	Feature Name	Feature Description
		Huawei's hierarchical QoS technology supports refined identification of user services, ensuring that voice streams are not lost and video streams are not frozen in the case of burst interference and heavy load
	Big-data-based intelligent optimization	Huawei big-data-based intelligent optimization technology is the latest combination of artificial intelligence (AI) and radio calibration technologies. It can adjust and optimize the network iteratively based on historical data, such as interference and load, to achieve the optimal configuration of the entire network
	High reliability	To prevent network unavailability caused by device faults, you need to deploy wireless access controller (WAC) hot standby
	WLAN wireless security	Rogue STAs and APs can be identified and countered. Attack identification and prevention are supported
Multinetwork convergence	Asset management (optional)	The locations of valuable assets in an enterprise can be tracked and managed
	Indoor navigation (optional)	Positioning and navigation services are provided for enterprise guests
User access authentication and policy control	User access and authentication	Multiple user authentication modes are supported, including 802.1X authentication for employees, portal authentication for guests, MAC address authentication for office devices, and different service set identifiers for guests and employees
	User permission control	Different rights control and access modes are configured for employees and guests. For example, guests have only Internet access rights, and employees can access the enterprise data center or office system

1. High-quality wireless network: The office network is the production network of an enterprise. A network's efficiency directly affects production efficiency. The network must be future-oriented and must be able to carry key enterprise services for the next 3–5 years. The wireless network must meet the following requirements:

   - Provides a large capacity of 100 Mbps anytime and anywhere. Key standards are Wi-Fi 6, and key technologies include MU-MIMO that supports multiple spatial streams (8 × 8).

- Provides anti-interference capabilities to ensure continuous networking for multiple APs, and supports continuous networking of 40/80 MHz bandwidth. Key technologies include smart antenna, dynamic anti-interference technology, and basic service set (BSS) color.

- Provides roaming capabilities. The key technology is 802.11k, 802.11v, and 802.11r smart roaming, as it optimizes the compatibility of roaming behaviors of mainstream STAs.

- Provides application identification and QoS capabilities. The key technologies include accurate application identification and hierarchical QoS.

2. Multinetwork convergence: The solution supports Internet of things (IoT) convergence. IoT APs can implement asset management based on radio frequency identification (RFID) and Bluetooth technologies, and implement indoor positioning based on Wi-Fi and Bluetooth technologies.

3. User access authentication and policy control: The solution supports enterprise-level user access authentication and provides comprehensive policy control capabilities.

9.1.3 Network Planning

1. Network coverage design

   Enterprise offices are usually an open area with several obstacles such as cubicles and pillars. APs are generally deployed 10–15 m apart to meet coverage requirements.

2. Network capacity design

   Enterprise office services increase rapidly. When considering network requirements in the next 3–5 years, it is recommended that APs supporting Wi-Fi 6 be used to provide a maximum of 12 spatial streams. In addition, enterprises' wireless STAs need to be upgraded to support Wi-Fi 6 as soon as possible to maximize the WLAN performance.

   APs that comply with Wi-Fi 6 have significantly improved capacity, multiuser concurrency, and anti-interference capabilities, as illustrated in Table 9.2. When such APs are used, the network planning is different.

TABLE 9.2   Performance Comparison of APs Supporting Wi-Fi 5 and Wi-Fi 6

	AP Performance Parameters	
Item	AP Supporting Wi-Fi 5	AP Supporting Wi-Fi 6
Channel bandwidth	20/40 MHz	40/80 MHz
Average STA throughput	50 Mbps	100 Mbps
Peak STA throughput	100 Mbps	300 Mbps
Number of concurrent STAs on a single frequency band	6–10	20–40

As illustrated in Table 9.2, when APs supporting Wi-Fi 6 are used, the peak STA throughput increases greatly. However, as most STAs on the live network do not support Wi-Fi 6, the performance of the entire network cannot be fully utilized even if APs supporting Wi-Fi 6 are deployed. In addition, it is estimated that the proportion of STAs supporting Wi-Fi 6 will increase to over 30% by the end of 2020, and network performance will gradually improve.

3. AP deployment design

In a discontinuous semiopen space with a small width, the distance between APs is 10–15 m, as illustrated in Figure 9.2.

In a continuous semiopen space with a large width, APs are deployed in W-shaped positions with spacing of 10–15 m, as illustrated in Figure 9.3.

In enterprise offices with few obstacles and a medium personnel density, it is recommended that settled APs with omnidirectional antennas be used and mounted on ceilings.

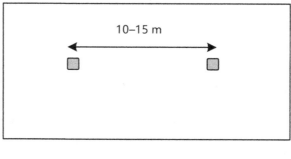

□AP position

FIGURE 9.2   AP deployment solution in enterprise offices (discontinuous semi-open spaces).

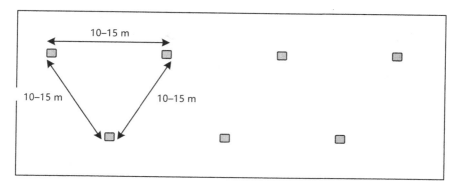

□ AP position

FIGURE 9.3 AP deployment solution in enterprise office scenarios (continuous semiopen spaces).

In addition, signal coverage may be required in corridors. To prevent indoor and outdoor signal interference, APs must be kept away from offices when being deployed in corridors. The distance between APs and solid masonry walls must be at least 3 m, and at least 6 m between APs and nonsolid masonry walls (such as gypsum boards and glass).

9.2 STADIUMS/EXHIBITION HALLS

Stadiums, exhibition halls, and large conference halls are typical high-density scenarios that have high requirements for network planning.

9.2.1 Business Characteristics

1. Space characteristics

Stadiums and exhibition halls have the following key characteristics (Figure 9.4):

- Space: Their shapes are unusual (e.g., most stadium stands have irregular curved surfaces) and their setting is diverse, often with ceilings of more than 10 m in height. As such, this environment greatly impacts the network planning solution and requires professional network planning and design.

- Blocking: Signal blockage is rare in stadiums but is severe in exhibition halls due to the deployment of exhibition booths.

FIGURE 9.4   An exhibition hall.

- Interference: Stadiums or exhibition halls have high ceiling height, and co-channel interference easily occurs when APs are deployed. Dense population in these venues indicates high STA density, leading to severe conflicts and interference between STAs.

2. Service characteristics

In stadiums and exhibition halls, WLANs primarily carry services for both personal and exhibition use.

- For personal services, user experience must be stable and the experienced rate must be higher than 16 Mbps to avoid experience deterioration (e.g., sudden rate decrease).

- Exhibition services may include professional live broadcast and video transmission; therefore, it is recommended that dedicated VIP areas be planned to deliver target services.

9.2.2 Best Practices of Network Design

Based on the service characteristics of stadiums and exhibition halls, the best practices of network design are as follows:

1. High-quality wireless network: Stadiums and exhibition halls are typical high-density deployment scenarios. The wireless network must meet the following requirements (Table 9.3):

TABLE 9.3   Feature List for Stadiums and Exhibition Halls

Type	Feature Name	Feature Description
High-quality wireless network	DFBS	APs support multiple modes, such as dual-radio, dual-radio + scan, and triple-radio, and can automatically switch between these modes to improve throughput in the coverage area of multiple scenarios
	Wi-Fi 6 OFDMA	The concurrency performance of Wi-Fi 6 is four times higher than that of Wi-Fi 5
	Narrow-angle directional antenna	Narrow-angle directional antennas are used to provide signal coverage in specified directions and reduce the interference with other APs APs with built-in narrow-angle directional antennas reduce installation costs and improve O&M efficiency
	MU-MIMO	As the number of spatial streams (one or two) of an STA is smaller than that of an AP, a single STA cannot fully utilize the performance of an AP, and so MU-MIMO technology is introduced to allow multiple STAs to access one AP for data transmission. After MU-MIMO enhancement is enabled, the capacity is at least doubled
	Load balancing	When a large number of STAs access the network, load balancing needs to be performed between APs, and between 2.4 and 5 GHz radios of the same AP
Multinetwork convergence	Asset management (optional)	The asset and personnel tracking and management functions are provided for exhibition enterprises
	Indoor navigation (optional)	Positioning and navigation services are provided for exhibition attendees

- Large capacity and high concurrency. One AP provides network access for at least 100 STAs with a data rate of 16 Mbps anytime and anywhere. The key technologies are orthogonal frequency division multiple access (OFDMA) and Dynamic Frequency Band Selection (DFBS) in Wi-Fi 6 (APs can automatically switch between the triple-radio and dual-radio modes).

- Anti-interference capability to ensure continuous networking for multiple APs. The key technologies are narrow-angle directional antennas (external or built-in), dynamic anti-interference technology, and BSS color.

2. Multinetwork convergence: The solution supports IoT convergence. IoT APs can manage asset labels and personnel nameplates based on RFID and Bluetooth technologies, and implement indoor positioning based on Wi-Fi and Bluetooth technologies.

9.2.3 Network Planning

This section uses the stadium stand as an example to describe the network planning in this scenario.

According to the service characteristics of stadiums, the bandwidth of a single user is 16 Mbps, and a dual-radio AP serves about 100 users.

The ceilings are high in stadiums, and to prevent interference between APs, it is advised to use APs with built-in narrow-angle directional antennas.

The network planning solution for the stadium stand includes wall- and ceiling-mounted scenarios.

1. 2.4 GHz wall-mounted scenario: As illustrated in Figure 9.5, APs are installed on the wall behind the last row of seats. It is recommended that external antennas with an angle of 18° (2.4 GHz) be used, and the distance between them be greater than 12 m. In the stadium stand, AP channels are periodically repeated in the sequence of 1, 9, 5, and 13.

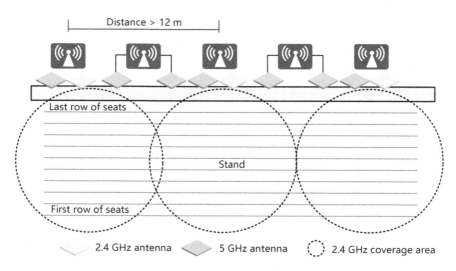

FIGURE 9.5   Network planning solution for 2.4 GHz wall-mounted scenario.

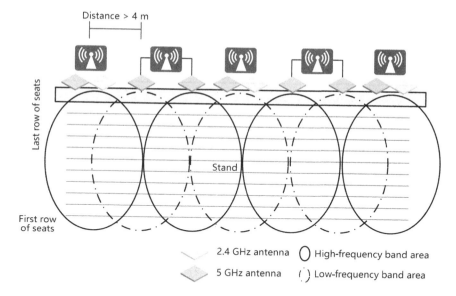

FIGURE 9.6   Network planning solution 1 for 5 GHz wall-mounted scenario.

2. 5 GHz wall-mounted scenario: As illustrated in Figure 9.6, external antennas with an angle of 15° (5 GHz) are recommended. The distance between 5 GHz antennas must be greater than 4 m to avoid interference between APs. A combination of high- and low-frequency bands is recommended. For example, in China's country code, channels 149–165 are recommended in high-frequency bands, and 36–64 in low-frequency bands.

In this solution, triple-radio APs with built-in narrow-angle directional antennas (with an angle smaller than 30°) can also be used. Figure 9.7 illustrates the coverage solution, where the distance between APs must be greater than 8 m, and if the 2.4 GHz radio is enabled, the distance between APs enabled with 2.4 GHz must be greater than 16 m.

3. Ceiling-mounted scenario

As illustrated in Figure 9.8, when the ceiling height is less than 20 m, APs are installed on the ceiling berms. It is recommended that external antennas with an angle of 18° (2.4 GHz) or 15° (5 GHz) be used. 2.4 GHz antennas must be deployed with spacing greater than 12 m and placed in the sequence of channels 1, 9, 5, and 13. 5 GHz antennas must be deployed with spacing of 4 m.

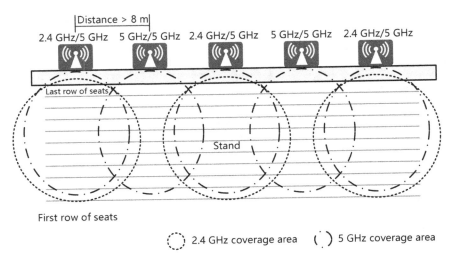

FIGURE 9.7 Network planning solution 2 for 5 GHz wall-mounted scenario.

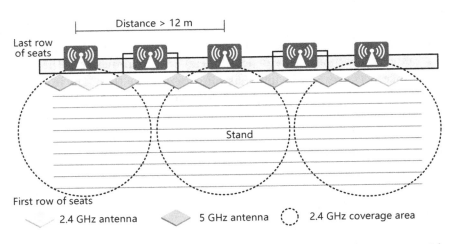

FIGURE 9.8 Network planning solution for ceiling-mounted scenarios with height of less than 20 m.

As illustrated in Figure 9.9, when the ceiling height is greater than 20 m, APs are installed on the berm of the ceiling. It is recommended that external antennas with an angle of 18° (2.4 GHz) or 15° (5 GHz) be used. 2.4 GHz antennas must be deployed with spacing of greater than 16 m (As the AP installation height increases, it also increases the overlapping coverage area of neighboring APs and causes severe

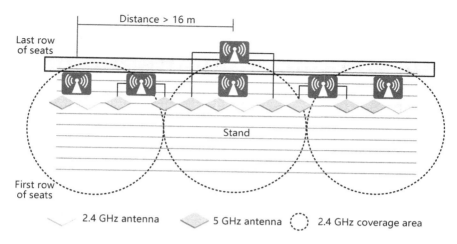

FIGURE 9.9 Network planning solution for ceiling-mounted scenarios with height of greater than 20 m.

interference. As a result, there needs to be more distance between APs.) and placed in the sequence of channels 1, 9, 5, and 13. 5 GHz antennas must be deployed with spacing of 4 m, meaning that 5 m feeders need to be purchased.

4. Channel planning and design in the stand area

Proper channel planning for each AP is crucial to WLAN performance, as it prevents interference from the network and existing channels. The general principle of channel planning is to keep the distance between intra-frequency APs and adjacent-frequency APs as long as possible to improve the channel reuse rate. In addition, channel planning for APs at the same stand tier and the upper and lower tiers also needs to be considered. Figure 9.10 illustrates the channel planning.

9.3 HOTEL ROOMS, DORMITORIES, AND HOSPITAL WARDS

Hotels, dormitories, and hospital wards typically have a large number of rooms in one building. In these places, rooms are densely distributed and each room has a small number of terminals. The wireless network mainly provides basic Internet access and mobile healthcare services in wards. Such scenarios tend to require deployment costs to be kept low.

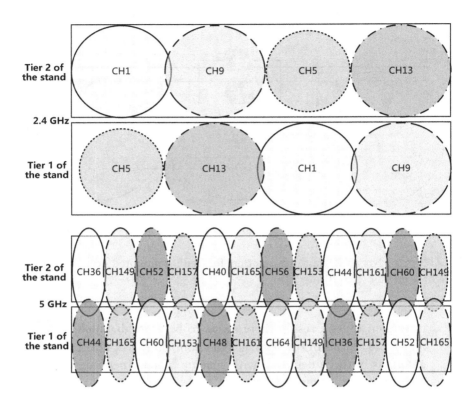

FIGURE 9.10 Channel planning for a two-tier stadium stand.

9.3.1 Business Characteristics

1. Space characteristics

Figure 9.11 shows a hotel room, dormitory, or ward scenario, which has the following key features:

- Space: The ceiling height is less than or equal to 4 m, and the area is about 20 m².

- Blocking: Major obstacles are bathroom walls.

- Interference: Microwave ovens and personal routers may be found in dormitories.

- Density: There are many rooms in one building, with a small number of STAs in each room.

FIGURE 9.11   Dense-room scenario.

2. Service characteristics

In hotel rooms and dormitories, personal Internet access services are usually provided, and the single-user experienced rate must reach 100 Mbps or higher. Special service assurance is required for personal entertainment applications, such as games and live broadcast.

In wards, mobile healthcare services must be considered. If customers require zero roaming of medical mobile terminals, the WLAN must support single frequency network (SFN) roaming.

9.3.2 Best Practices of Network Design

Based on the service characteristics of dense-room scenarios, the best practices of network design are as follows. Table 9.4 lists the involved features.

High-quality wireless network: Hotel rooms, dormitories, and wards are typical room-intensive scenarios. In these places, the wireless network must meet the following requirements:

- Provides a large capacity of 100 Mbps anytime and anywhere. The key technologies (standards) are Wi-Fi 6 standards, smart antenna, agile distributed architecture, and SFN roaming.

- Provides application identification and QoS capabilities. The key technologies are accurate application identification, hierarchical QoS, and game acceleration.

TABLE 9.4   Feature List of High-Quality Wireless Networks in Dense-Room Scenarios

Feature Name	Feature Description
AP performance	The AP complies with Wi-Fi 6 standards and supports six spatial streams (2.4 GHz 2×2 MIMO and 5 GHz 4×4 MIMO). Multiple spatial streams can meet the large-capacity requirement of 100 Mbps for a single user
Smart antennas	Huawei smart antenna technology increases the AP coverage by approximately 15%, eliminates coverage holes in rooms, and reduces interference to other APs and STAs through beamforming
MU-MIMO	As the number of spatial streams (one or two) of an STA is smaller than that of an AP, a single STA cannot fully utilize the performance of an AP, and the MU-MIMO technology is introduced to allow multiple STAs to access one AP for data transmission After MU-MIMO enhancement is enabled, the capacity can be doubled at least
Smart roaming	This feature provides good roaming experience for users when moving to rooms or corridors
Service-based QoS control (game acceleration)	Hotel rooms, dormitories, and wards mainly provide personal Internet access. Games are most sensitive to latency. Huawei QoS management is specially optimized for games to ensure that the game latency is less than 15 ms
SFN roaming	In hospital wards, mobile healthcare services are sensitive to roaming delay. The SFN roaming feature provides seamless roaming to ensure high-quality experience of mobile healthcare services
Agile distributed architecture	In scenarios with a large number of rooms on each floor, WLAN construction needs to strike a balance between deployment costs and user experience. Deploying an independent AP in each room greatly increases deployment costs and complexity. The agile distributed architecture using a central AP and RUs can solve this problem
Wired access	The AP provides downlink wired interfaces, allowing wired terminals to connect to the network through network cables. The downlink wired interface can authenticate access STAs to ensure network security

9.3.3 Network Planning

Hotel rooms, dormitories, and wards are densely distributed, with each room usually having fewer than ten users. It is recommended that one remote unit (RU) be deployed in each room to meet capacity and coverage requirements.

Figure 9.12 illustrates the network planning for hotel rooms.

It is recommended that RUs be deployed in rooms. Figure 9.12 illustrates the RU deployment positions. One RU is placed in each room on the desk

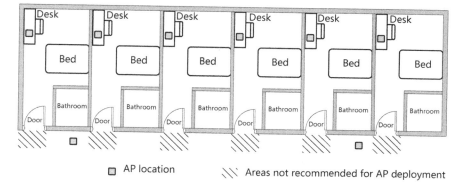

□ AP location   \\\ Areas not recommended for AP deployment

FIGURE 9.12   Agile distributed network planning solution.

(wall-mounted in dormitories and wards, and the installation position can be the wall where the door is located). Further, an AP should be deployed outside every three rooms in the corridor, away from room doors.

9.4 CLASSROOMS

There are two types of classroom scenarios. The first involves scenarios where only teaching services are provided, for example, classrooms in primary and secondary schools. The other involves both teaching services and Internet access services, for example, classrooms and lecture halls in colleges and universities.

9.4.1 Business Characteristics

1. Space characteristics

    Figure 9.13 illustrates the classroom scenario, which has the following key features:

    • The ceiling height is less than or equal to 4 m, while its total area is approximately 80 m².

    • No obstacles exist, and there is limited signal interference.

2. Service characteristics

    There are different service requirements in primary and secondary school classrooms compared with the classrooms and lecture halls of colleges and universities. Specifically, primary and secondary school classrooms mainly provide teaching services, such as video, electronic whiteboard, file download, and desktop sharing, whereas

FIGURE 9.13   Classroom scenario.

colleges and universities also provide web browsing and social networking applications

However, the two scenarios both have a large number of concurrent users. In particular, the students in classrooms of primary and secondary schools may concurrently play a course video during their class, for example.

9.4.2 Best Practices of Network Design

Based on the service characteristics of classroom scenarios, the best practices of network design are as follows. Table 9.5 lists the involved features.

High-quality wireless network: Classroom scenarios feature large capacity, high concurrency, and multiple users. The classroom's wireless network must meet the following requirements:

- Provides a large capacity of 100 Mbps anytime and anywhere. It must employ Wi-Fi 6 standards and key technologies, such as MU-MIMO that supports multiple spatial streams (8 × 8) and DFBS.

- Provides anti-interference capabilities to ensure continuous networking for multiple APs and supports continuous networking of 80/160 MHz bandwidth. Its key technologies include smart antenna, dynamic anti-interference technology, and BSS color.

TABLE 9.5   Feature List of High-Quality Wireless Networks in Classroom Scenarios

Feature Name	Feature Description
AP performance	The AP complies with Wi-Fi 6 standards and supports 12 spatial streams. Multiple spatial streams meet the large-capacity requirement of 100 Mbps for a single user
DFBS	APs support dual-radio (4+8) and triple-radio (4+4+4) modes and automatically switch between the two modes to improve the wireless throughput in the coverage area of multiple scenarios
Wi-Fi 6 OFDMA	The concurrency performance of Wi-Fi 6 is four times higher than Wi-Fi 5
MU-MIMO	As the number of spatial streams (one or two) of an STA is less than that of an AP, a single STA cannot fully utilize the performance of an AP, and MU-MIMO technology is introduced to allow multiple STAs to access one AP for data transmission After MU-MIMO enhancement is enabled, capacity can be doubled
Smart antenna	Huawei smart antenna technology increases AP coverage by approximately 15% and reduces interference to other APs and STAs through beamforming
Load balancing	In most lecture hall scenarios, two to three APs are deployed. In this case, load balancing needs to be implemented among the APs to prevent a large number of STAs from connecting to the AP closest to the entrance when users enter the hall
Anti-interference feature	In classrooms separated by walls, the dynamic anti-interference technology and the Wi-Fi 6 BSS color technology are used to implement continuous networking with a large bandwidth of 80 or 160 MHz

9.4.3 Network Planning

Figure 9.14 illustrates the network planning for classrooms with high requirements on appearance, capacity, and signal coverage.

- Common classrooms: In rooms with an area smaller than 100 m², one AP is mounted on a high beam or ceiling.

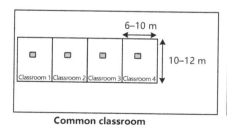

Common classroom

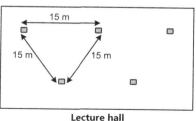

Lecture hall

☐ AP location

FIGURE 9.14   Network planning solution for classrooms.

- Lecture halls: Multiple APs are installed on the ceiling in a W-shaped layout with 15 m distance between them. Each AP is used to provide access for 60 STAs.

9.5 SHOPPING MALLS, SUPERMARKETS, AND RETAIL CHAINS

In shopping malls, supermarkets, and retail chains, WLANs were originally deployed as value-added services to attract customers and improve customer satisfaction. However, WLANs are now used to carry new services, such as wireless cashier, electronic shelf labeling, and precision marketing.

9.5.1 Business Characteristics

1. Space characteristics

    Figure 9.15 shows the shopping mall, supermarket, and retail chain scenarios, which has the following key features:

    - Space: A shopping mall and supermarket's building height and area vary greatly. Large shopping malls usually have large open areas (such as atriums) that may cause inter-floor interference. The area of a retail chain store ranges from 10 to 100 m².

    - Blocking: The blocking area is large, as it includes product shelves, walls, and glass doors.

    - Interference: Wi-Fi interference is experienced from the same floor or other floors, and non-Wi-Fi interference is experienced from wireless communication technologies that use frequency bands that overlap with Wi-Fi, such as Bluetooth.

FIGURE 9.15 Shopping malls, supermarkets, and retail chains.

2. Service characteristics

WLAN services include enterprise applications and individual applications.

- Enterprise applications: An enterprise's internal applications, wireless payment applications, and IoT applications.

- Individual applications: Web browsing, such as commodity price comparison and shopping navigation; social and game applications, such as voice applications and mobile games.

9.5.2 Best Practices of Network Design

Based on the service characteristics of shopping malls, supermarkets, and retail chains, Table 9.6 lists the involved features in the best practices of network design.

1. Multinetwork convergence: This solution supports IoT convergence. That is, IoT APs are used to manage asset labels based on RFID, and implement indoor positioning based on Wi-Fi and Bluetooth technologies.

2. Network management capability: This solution supports multi-branch cloud management and IPsec tunnel functions.

9.5.3 Network Planning

The enterprise services used in shopping malls and supermarkets do not require high bandwidth. Instead, their primary service requirements are signal coverage, latency, and reliability. For example, the interference between WLAN and IoT (such as electronic shelf labels (ESLs) and BLE labels) needs to be avoided. In these scenarios, personal applications also do not require a high bandwidth, and signal coverage is a top concern due to many obstacles and complex building structures. Therefore, the service rate of 16 Mbps can be used to meet service requirements.

Key coverage scenarios to be considered include public areas, corridors, supermarkets, and retail chains.

1. Network planning solution for the public area/corridor scenario

As illustrated in Figure 9.16, settled APs with omnidirectional antennas are deployed below the ceiling. The distance between APs is 20–25 m.

TABLE 9.6   Features in the Shopping Mall and Super Market Scenario

Type	Feature Name	Feature Description
Network management	Security tunnel function	For franchised stores that need to remotely access their headquarters' servers, virtual private network security tunnel functions, such as Internet Protocol Security (IPsec) and Secure Sockets Layer (SSL), are used to ensure service security
	Cloud management	For small- and micro-branches without professional IT O&M personnel, a cloud management architecture is recommended. Network deployment can be completed without professional skills, and service providers can provide technical assurance
	WAN authentication escape	If a WAC is deployed in the headquarters to globally manage APs in branches, the APs in branches can still provide Internet access services if the WAC fails
	Wired access	An AP provides downlink wired interfaces to allow wired terminals to connect to the network through network cables. The downlink wired interface can authenticate access STAs to ensure network security
Multinetwork convergence	IoT AP	ESLs effectively reduce costs, quickly adjust commodity prices and information, and improve customers' shopping experience. In addition, asset labels are used to effectively manage the valuable assets of enterprises, improve asset stocktaking efficiency, and monitor asset locations in real time. What's more, indoor positioning is usually used in large shopping malls and supermarkets to provide customers with personalized, value-added services such as shopping guides and car searching services in parking lots, improving customers' shopping experience

2. Network planning solution for supermarket scenarios

As illustrated in Figure 9.17, settled APs with omnidirectional antennas are deployed.

- Common area: APs are installed on the ceiling in an equilateral triangle formation, with a 20 m distance to each AP.

- Hollowed-out area on the top of the ceiling frame: APs are installed on the ceiling frame with a distance of approximately 20 m to each AP.

3. Network planning solution for retail chain scenarios

Retail chain stores include retail stores and cafes.

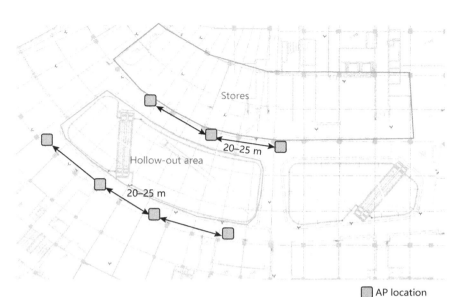

FIGURE 9.16   Network planning solution for the public area/corridor scenario.

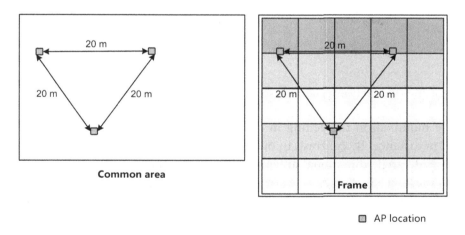

FIGURE 9.17   Network planning solution for the supermarket scenario.

The network planning solution for retail stores is illustrated in Figure 9.18. Large stores need to deploy APs with a coverage radius of 8–10 m, while small stores each need to deploy one AP. The coverage area of a single AP is $100–200\,m^2$.

Figure 9.19 illustrates the network planning solution for the cafe scenario. In indoor scenarios, APs with built-in omnidirectional antennas

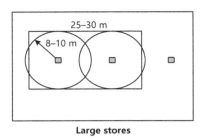 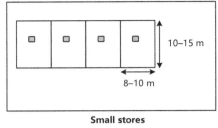

Large stores   **Small stores**

□ AP location

FIGURE 9.18   Network planning solution for the retail store scenario.

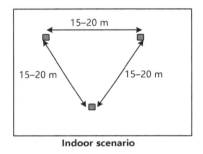 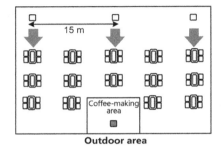

Indoor scenario   **Outdoor area**

■ Indoor AP location
□ Outdoor AP location

FIGURE 9.19   Network planning solution for the cafe scenario.

are mounted on the ceiling in an equilateral triangle formation with a 12 m distance. By contrast, in outdoor scenarios, outdoor APs with built-in wide-angle directional antennas are recommended. APs are deployed on the long edges of the cafe area (wall-mounted or pole-mounted) to provide coverage for users in the seating area. For details, refer to outdoor AP deployment illustrated in the following figure. In cafe scenarios, the coffee-making area is usually an independent indoor structure. Therefore, it is recommended that an AP with built-in omnidirectional antennas be installed on the ceiling to cover this area.

9.6 SQUARES AND STREETS

Squares and streets are two types of scenarios that both feature wide coverage areas and harsh outdoor environments. More specifically, squares refer to large open areas, such as public squares, playgrounds, parks, and scenic spots. Streets, on the other hand, refer to long narrow areas, including streets for motor vehicles and pedestrian walking streets.

9.6.1 Business Characteristics

1. Space characteristics

     As illustrated in Figure 9.20, square and street scenarios include the following key features:

     • Space: Both involve outdoor areas with either wide coverage or narrow and long coverage.

     • Blocking: Both scenarios feature many obstacles, such as buildings and trees, which place significant challenges on network planning.

     • Interference: Both experience interference from radar signals in the same frequency band.

     • Environment: Both experience harsh outdoor environments, such as salt mist, thunderstorms, heavy rain, and sandstorms.

2. Service characteristics

     While traveling through squares and streets, users mainly use Internet services such as web browsing and social networking applications.

9.6.2 Best Practices of Network Design

In squares and streets, outdoor APs are deployed to meet the technical specifications of outdoor scenarios. Table 9.7 lists the features involved in squares and streets.

FIGURE 9.20   Square/street scenario.

TABLE 9.7   Square/Street Scenario Feature List

Type	Feature Name	Feature Description
Outdoor AP performance	Dustproof and waterproof	To cope with complex outdoor environments, outdoor APs with dustproof and waterproof capabilities are required
	Surge protection	To protect APs against power surges caused by lightning, outdoor APs must have surge protection capabilities
	Anticorrosion	Outdoor APs can work in coastal or other corrosive environments
	Wide temperature	Outdoor temperatures vary greatly throughout the year, requiring APs to support a wider range of operating temperature
	Directional antenna	Directional antennas are used to extend the coverage distance or improve backhaul capability with high antenna gains
Wireless cascading capability	Mesh/wireless distribution system (WDS)	Mesh or WDS allows APs to set up wireless links between them to transmit data, especially in environments where cables are difficult to route. AP wireless cascading enables flexible networking and reduces costs. This solution is mainly used for wireless backhaul of video surveillance and network coverage in rural areas
	Outdoor long-distance coverage capability	Outdoor APs can work with directional antennas to provide 300–800 m ultra-long-distance coverage, which is applicable to wide coverage in rural areas

9.6.3 Network Planning

In squares and streets, each user is allocated approximately 8–16 Mbps of bandwidth to access the Internet with their mobile phones.

Table 9.8 describes outdoor scenarios.

1. Network planning solutions for squares, parks, and tourist attractions
    APs and antennas are flexibly selected based on factors such as the coverage area, site location, and capacity.

    In different scenarios, different coverage rules need to be applied. For scenarios with a wide view and few obstacles, as illustrated in Figure 8.10, outdoor APs connecting to omnidirectional antennas are used, while each AP should be separated by a distance of between 50 and 60 m.

    For squares in front of a building, it is recommended that wall-mounted APs with wide-angle directional antennas be used.

TABLE 9.8   Outdoor Scenarios

Scenario Type	Building Characteristics	User Characteristics	Network Characteristics	Product Selection
Squares, parks, and tourist attractions	Open areas	Mainly mobile phone users, with high mobility and low bandwidth requirements	Wide coverage and low bandwidth requirements	If APs can be deployed in a central area, deploy APs with omnidirectional antennas. Otherwise, deploy APs with directional antennas on the top of tall buildings
Commercial/ pedestrian streets	Building clusters with varied heights	Mainly mobile phone users, with high mobility and low bandwidth requirements	Road coverage and low bandwidth requirements	Use wide-angle antennas to expand the coverage areas and deploy APs in the W-shaped layout along the streets

The recommended installation height for these APs is 4–6 m, while the recommended distance between APs is 30–40 m.

2. Commercial/pedestrian streets

Commercial/pedestrian streets can be divided into either wide green streets or nongreen streets. In nongreen streets, the coverage solution is applicable to narrow streets without any greenery or blocked signals, as illustrated in Figure 9.21. The following needs to be considered when deploying APs on streets:

- Deploy APs with directional antennas in the W-shaped layout along a street, and ensure that channels do not overlap.

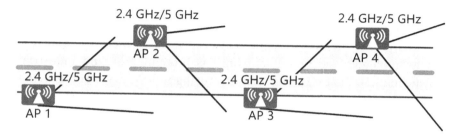

FIGURE 9.21   Coverage solution for nongreen streets.

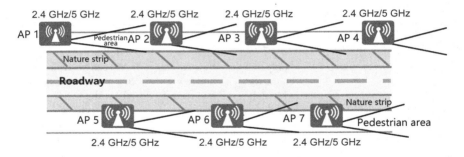

FIGURE 9.22   Coverage solution for wide green streets.

- Install APs on the same side of the street if installation conditions are restricted.

- Install APs at a height of between 6 and 8 m if there are areas with low-lying greenery along the street.

By contrast, the wide green street coverage solution is applicable to streets with large strips of greenery and forestry, as illustrated in Figure 9.22. In this solution, the following needs to be considered when deploying APs on streets:

- If the pedestrian area needs to be covered, APs must be deployed on the pedestrian area of both sides of the street.

- If both pedestrian areas and roadways need to be covered, additional APs need to be installed on the roadway, and antennas need to be installed on the side with greenery. For more details, see the deployment in nongreen scenarios.

- AP antennas must be installed above the greenery to prevent it from blocking signals.

9.7 PRODUCTION WORKSHOPS AND WAREHOUSES

In Industry 4.0, the deep integration between next-generation information technologies and manufacturing has brought profound industrial transformation. That is, in the manufacturing industry, Wi-Fi 6 is being employed to build wireless networks with high-performance and high-reliability to facilitate intelligent transformation.

However, as most of the processing equipment used in production workshops is metal-based, their large size and dense distribution may cause coverage holes in WLANs. In addition, during the product test, ultra-large firmware needs to be downloaded in a short time. This requires that the wireless network provides stable and reliable connections, as well as large bandwidth. The production workshop network solution design's objectives are to deliver the workshop with large capacity, high stability, and strong coverage.

In addition, the warehouse environment, due to its complex nature and dense distribution of obstacles, may also cause wireless network coverage holes. The warehouse wireless network solution aims to provide a stable and reliable wireless network with no coverage holes and seamless roaming.

9.7.1 Business Characteristics

1. Space characteristics

   The production workshop scenario, illustrated in Figure 9.23, has the following key features:

   - Space: An indoor environment with a large amount of densely distributed metal devices and equipment.

   - Blocking: Signals are prone to being blocked and coverage holes frequently occur, posing high requirements on network planning and design.

   - Interference: Co-channel or adjacent-channel interference is likely to occur due to limited spectrum resources.

FIGURE 9.23   Production workshop scenario.

FIGURE 9.24   Warehouse scenario.

In Figure 9.24, the warehouse scenario is illustrated to include the following key features:

- Space: an indoor area with complex layout and dense shelves.

- Blocking: The indoor area includes many obstacles, as well as many shelves that are densely located together, posing high requirements on network planning and design.

- Interference: The area's network frequently experiences interference from other wireless systems, such as IoT systems.

- Environment: In some cases, warehouses have special environment requirements, for example, low-temperature warehouses that keep a large amount of frozen goods.

2. Service characteristics

- Production workshop: Data transmission of wireless customer-premises equipment (CPE) is provided. Specifically, the firmware upgrade on the head unit requires the bandwidth of a single terminal to be 300 Mbps, and HD inspection video collection, such as 4K video transmission, requires a bandwidth of 25 Mbps. IoT technology-based sensors and instruments used for communication may also be located in the production workshop.

- Warehousing: Automated guided vehicles (AGVs) need to send and receive instructions and report their task status during movement. To do this, they require wireless networks that feature zero packet loss and low latency (less than 50 ms). In addition, the warehouse's PDA scanner needs to collect goods information and transmit the information to the data center using the wireless network. To perform this, the PDA scanner requires uninterrupted data transmission during PDA roaming.

9.7.2 Best Practices of Network Design

Table 9.9 lists the recommended features to be employed during network design based on the service characteristics of the production workshop and warehouse scenario.

9.7.3 Network Planning

In the production workshop scenario, the main service requirements include the data transmission of wireless CPEs and IoT devices, such as sensors and instruments.

1. Wireless CPE transmission scenario

In traditional networks, large data packets were transmitted through wired connections, which required complex cable routing. However, most scenarios now complete data transmission through wireless connections to improve and simplify the workshop environment. For example, the firmware upgrade of head units requires large bandwidth and high reliability to avoid batch shutdown, and can now be simply and flexibly completed through wireless connections.

TABLE 9.9   Features in the Production Workshop and Warehouse Scenarios

Feature Name	Feature Description
IoT AP	Based on WLAN, this solution implements co-site, co-backhaul, and unified entrance and management of various IoT connection modes on APs. This solution is flexible and scalable
WLAN security	Rogue STAs and APs can be identified and prevented from accessing the network. It supports attack identification and prevention
	It supports spectrum scanning to determine the interference source in the environment
Automatic navigation roaming optimization	AGVs can efficiently search for Wi-Fi signals and quickly determine the target AP for roaming, achieving zero packet loss and low latency during roaming

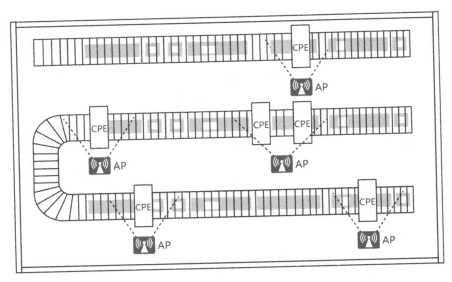

FIGURE 9.25   Wireless CPE transmission network planning solution.

To meet bandwidth and reliability requirements, it is recommended that Wi-Fi 6-capable APs with directional antennas be used. It is also recommended that each AP be connected to a CPE, and the AP installation height be less than 8 m. In addition, the coverage direction of directional antennas must be aligned with the CPE, and there must be no obstacles blocking the AP and the CPE. Strong interfering signals must also not exist around the APs to ensure a high transmission rate over the air interface, as illustrated in Figure 9.25.

2. Coexistence of IoT devices

Many IoT devices, such as sensors and instruments, are found in the production workshop. IoT devices transmit data using short-range wireless communications technologies such as ZigBee and WirelessHART, while they also meet the reliable, stable, and secure wireless communication requirements of real-time factory applications in the industry.

These devices are usually installed in specific areas within a production workshop. For example, an IoT AP may be installed in a specific position based on the positions of other IoT devices, ensuring data transmission is reliable and stable. However, as the transmission distances vary between different IoT devices, the installation heights and spacing requirements of APs are also different. It is recommended that the spacing between an AP and IoT device not exceed the maximum distance supported by an IoT protocol.

Typical warehouse applications include AGVs and PDA scanners, both of which do not require a high bandwidth and are sensitive to latency and packet losses. In this case, the network planning focus is on providing signal coverage and reliability, as well as reducing latency.

- Network planning solution for AGV scenarios

  Fixed-area AGVs: In this scenario, AGVs are used to move goods between different areas of warehouses. As the rack height is approximately 2.5 m, to ensure the reliability of wireless communication, the signal strength of an entire warehouse must be greater than −65 dBm.

  In AGV scenarios, it is recommended that APs with external omnidirectional antennas be installed on the warehouse ceiling. In addition, the APs should be installed above the aisle to reduce signal blocking between the APs and AGVs. In the warehouse, APs must be installed at a height of 6 m and deployed in an equilateral triangle layout (with spacing of 15 m), as illustrated in Figure 9.26.

  Cross-region AGVs: In this scenario, an AGV moves goods along a specified driving route, requiring signal strength to be greater than −65 dBm. Depending on its planned route, the AGV can move throughout the entire warehouse, and therefore

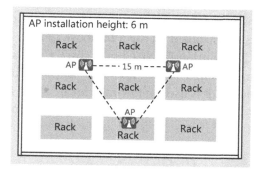

FIGURE 9.26   Network planning solution for AGV scenarios.

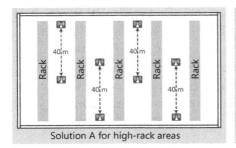
Solution A for high-rack areas

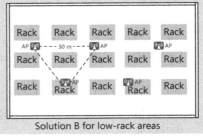
Solution B for low-rack areas

FIGURE 9.27 Network planning solution for the PDA scanner scenario.

seamless roaming must be ensured during network planning to avoid packet loss.

In this scenario, it is recommended that APs with an external omnidirectional antenna be installed on the ceiling above the driving route; however, the installation height should not exceed 6 m. In addition, APs should be deployed at an equal distance apart (15 m) on the driving route. Based on the signal strength in field test, APs could also be added at the warehouse edges to improve signal strength.

- Network planning solution for PDA scanner scenarios

This scenario includes the high-rack area, low-rack area, pickup area, and packing area. The high-rack area is used for pickup by forklift trucks, whereas the low-rack area and other areas are used for manual pickup. However, as signals are prone to being blocked in the high-rack areas, different network planning solutions are used for the high-rack and low-rack areas, as illustrated in Figure 9.27.

In this scenario, settled APs with external omnidirectional antennas are installed on the warehouse ceiling. As illustrated in Figure 9.27, Solution A is used in high-rack areas, and it involves deploying APs in a single aisle, each separated by 40 m. By contrast, Solution B is used in low-rack areas, and it involves deploying APs in an equilateral triangle, with a 30 m space between each AP.

CHAPTER 10

Enterprise WLAN O&M

NETWORK O&M INVOLVES THE essential network maintenance performed routinely by campus network administrators to ensure the normal and stable running of networks. Network O&M includes routine monitoring, network inspection, device upgrade, troubleshooting, and network change. More specifically, network change refers to the adjustment of network service configurations or the replacement of network devices; that is, it involves the design and deployment of small-scale services or devices. However, the design and deployment part has been explained in other chapters and will not be described here. Instead, this chapter will focus on the other abovementioned items.

10.1 ROUTINE MONITORING

To ensure network service quality, network administrators need to monitor the running indicators and device status to learn about the operation status of the entire network. If there is no strict network management requirement, network administrators can use the built-in web network management system (NMS) to perform basic monitoring. However, if there is, network administrators need to use a standalone NMS to periodically collect data for monitoring.

10.1.1 Monitoring Method

1. Local web-based NMS

    Wireless access controllers (WACs) and fat APs have built-in web network management functions. Network administrators can use the web network management function to monitor key device indicators. However, due to the limited storage space and data processing capability of the devices, the built-in Web NMS can only view real-time monitoring information, whereas the standalone NMS can store monitoring information for several months.

2. SNMP mode

The Simple Network Management Protocol (SNMP) released in 1990 is the management protocol used by traditional NMS to efficiently monitor and manage network devices in batches. In addition, SNMP can monitor network devices of different types and vendors in a unified manner. Traditional NMS obtains network monitoring data by receiving trap alarms reported by devices and periodically reading management information base (MIB) node data from devices through SNMP. For example, eSight is a network management device that is used in traditional campus networks to monitor and manage multiple network devices, including WACs, APs, and switches.

Traditional SNMP-based NMS has the following limitations:

- Traditional NMS is deployed on the user network and is costly for small-scale campus networks or campus networks with many branches. As a result, many networks run without an NMS, leading to frequent network faults and poor user experience.

- Traditional NMS monitors the network status from the perspective of devices and collects alarms, logs, command lines, or MIB data on devices to monitor network faults. However, many network faults are not device faults and cannot be detected from the device side. In this sense, the network needs to be monitored from the perspective of users and applications, so as to comprehensively detect network faults.

- Traditional NMS obtains operation data by proactively and periodically accessing devices. However, not only does this take a long time, but it also consumes valuable resources. A traditional NMS also cannot detect micro faults within minutes or shorter. Each time a large amount of data is obtained, the CPU usage of the device increases, impacting network services carried on the devices.

3. Telemetry mode

Telemetry is a next-generation network monitoring technology that, when compared with traditional NMS, includes the following improvements:

- Telemetry uses the cloud management mode. The network monitoring service is deployed on the cloud server to ensure that all devices are managed in a centralized manner. Its deployment and maintenance costs are low, while network administrators can monitor the network anytime and anywhere.

- Telemetry is used to monitor device, network, user, and application faults. Devices integrate network-, device-, user-, and application-level performance probes to detect the running status and the quality of each layer within minutes or even seconds. Then, the data is packed and sent to the server, which processes and integrates the data to generate a running track and trend chart of each application. By doing this, administrators are provided comprehensive data support for subsequent network fault analysis and location.

- Based on the raw data sent by telemetry, the server automatically evaluates the quality and converts the performance indicator data that is difficult to understand into quality scores, helping network administrators quickly identify the network running status.

- Telemetry provides big data storage and mining capabilities. That is, to collect a large amount of network device data and record data in seconds, the big data storage and query function are introduced on the server side to support the long-term storage of a large amount of network data. Historical network data generally needs to be stored for at least three months to more than half a year to support historical problem analysis and troubleshooting.

10.1.2 Major Monitoring Metric

Table 10.1 lists the device monitoring indicators that need to be considered when designing a typical wireless local area network (WLAN).

TABLE 10.1   Typical WLAN Monitoring Indicators

Indicator Category	Indicator Name		Description
Device information (WAC and AP)	Basic equipment information	Device name, model, hardware PCB, bill of materials (BOM), memory and storage space size, MAC address, and system time	Basic information about the device, which is generally static information
	Equipment status	Startup time, online duration, restart cause, CPU usage, memory usage, AP online status, AP online failure/offline cause, port working status, negotiated rate, and number of received and sent packets	Dynamic information about a device, which indicates the basic running status of the device. Generally, a device with abnormal indicators sends a trap alarm or performs some recovery actions For example, if the memory usage keeps increasing, a memory leak may occur. In this case, the device generates an alarm or even restarts to rectify the fault
RF and air interface information	RF information	Operating status, frequency band, channel, bandwidth, working mode, and RF transmit power	Basic RF specifications, in which the channel and RF transmit power requires special attention
	Air interface environment	Channel usage, packet loss rate, packet error rate, retransmission rate, co-channel interference strength, adjacent-channel interference strength, and non-Wi-Fi interference strength	These are important indicators of air interface quality. In the event of poor air interface quality, users may fail to obtain a good user experience. In this case, you need to check the interference source, adjust optimization parameters, or even rectify the radio network planning
STA information	Basic information	User MAC address, user name, STA type, authentication mode, associated AP/AP group, access service set identifier (SSID), online duration, online failure/offline cause, and application	These indicators are used for user data analysis, such as statistics on STA type proportions and user behavior analysis (e.g., stay duration and access application type distribution)

(Continued)

TABLE 10.1 (*Continued*)   Typical WLAN Monitoring Indicators

Indicator Category		Indicator Name	Description
	Key service indicators	Online success rate, online delay, frequency band, received signal strength indicator (RSSI), signal-to-noise ratio (SSID), negotiated rate, throughput, retransmission rate, packet loss rate, and roaming trajectory	Key indicators of the air interface can reflect whether air interface experience is smooth. In addition, the network load, user roaming status, and key algorithm running status can be analyzed. For example, network administrators can check whether the spectrum navigation function is normal based on indicators, such as the proportion of users using 5 GHz, radio frequency load, and signal strength
Service information	Service indicators	For example, the application, start time, end time, IP addresses of both parties, mean opinion score (MOS), jitter, delay, and packet loss rate of a voice or video session	Related service indicators can be used for observation in actual environment

10.2 NETWORK INSPECTION

Routine monitoring focuses on the running status of the current network and devices, while network inspection focuses on potential network issues that may impact services to detect and eliminate these risks before service deterioration. Network device vendors periodically release device precautions or rectification notices, and network administrators in the early development stages of network monitoring needed to log in to each device one by one through the management interface to check for the issues mentioned in the precautions or rectification notices. With increasingly large networks and diversified network device types, it is becoming more and more difficult to manually perform a comprehensive inspection on network devices.

The health check tool implements automatic inspection. Figures 10.1 and 10.2 show the GUIs of the eDesk, a network health check tool developed by Huawei.

FIGURE 10.1   Standalone eDesk GUI (1).

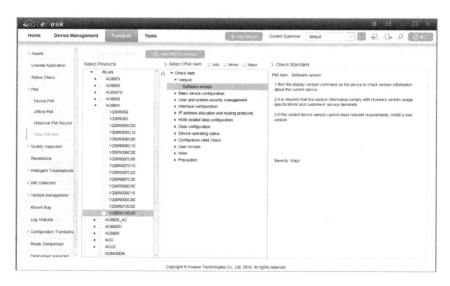

FIGURE 10.2   Standalone eDesk GUI (2).

Tool-based network inspection has the following advantages:

- The data collection and analysis processes are conducted through scripts to implement automatic analysis and identification of risks and problems, reducing skill requirement and workload.

- By automatically traversing all devices in batches, the eDesk implements automatic inspection of different types of devices on the entire network, reducing the inspection workload.

- Device vendors periodically release the latest tool version and scripts to accumulate inspection experience and ensure inspection quality.

- In addition to inspection, the eDesk also provides other functions, such as querying known bugs and applying for licenses.

Table 10.2 lists the inspection items.

TABLE 10.2   Inspection Items

Inspection Category	Inspected Item	Description
Version	Software version	Check whether the software and patch are officially released and whether the patch is correctly installed
Basic device configurations	Configuration file, system time, license, and debug switch	• Check whether the configuration file is saved properly and whether the settings and system time are correct • Check whether the license file is normal, whether the NMS service is enabled, and whether the debug function is disabled
	Route and VLAN check	Check whether a blackhole route is configured and whether the VLAN is correctly created
	Configuration check of common services such as authentication, Dynamic Host Configuration Protocol (DHCP), Virtual Router Redundancy Protocol (VRRP), channel, and optimization	Check whether common configurations are correct and proper
	Other service check	Check whether the configurations of noncommon services such as wireless distribution system (WDS), Mesh, and Wireless Intrusion Detection System (WIDS) are correct

(Continued)

TABLE 10.2 (*Continued*)   Inspection Items

Inspection Category	Inspected Item	Description
Device running status	Basic component status, important protocol and entry status, alarms, and service busy status	• Check the basic status of components such as the power supply, fan, temperature, CPU, memory, and storage space • Check the running status of basic protocols such as VRRP, Spanning Tree Protocol (STP), and Generic Routing Encapsulation (GRE)
	Basic component status, important protocol and entry status, alarms, and service busy status	• Check the Address Resolution Protocol (ARP) entries, Media Access Control (MAC) address learning, network address translation (NAT), central processor CAR (CPCAR), and traffic status • View major alarms
	Port status and statistics	Check whether there are idle ports that are not disabled, whether the port remarks correctly describe the usage, common port configurations, and port statistics (check whether there are error packets)
	Basic AP status	Check the AP online status, working mode, heartbeat, and upgrade parameters
	Whether the status of the server and device connected to the user is normal	Check the connectivity and configuration of servers such as the Remote Authentication Dial-In User Service (RADIUS) server, and check whether the number of access users reaches the limit of a certain access algorithm
	Air interface environment	Check the channel usage, 5 GHz usage, and wireless user throughput

The standalone eDesk inspection tool includes the following challenges:

- The inspection tool needs to be connected to the device network and the network administrator must carry the tool to the site for inspection.

- The inspection tool needs to be upgraded continuously to ensure the validity and comprehensiveness of the inspection result.

- The data collected by the inspection tool is used only for a single inspection. The data needs to be collected again for the next inspection, resulting in low data usage.

The next-generation network inspection technology addresses the preceding challenges.

Cloud tools are used for inspection. Devices can be inspected after being connected to the cloud management network, eliminating the need for on-site visits. The network administrator only needs to specify an inspection scope and time to automatically complete inspection on the cloud. This ensures that the inspection tool is always the latest version without being upgraded.

The inspection function uses the architecture that separates collection from analysis. The data collected during each inspection is saved on the server. You can add and analyze the collected data at any time to improve data usage, reduce dependency on devices, and analyze the time trend of the data collected multiple times.

For example, Huawei's cloud service platform ServiceTurbo Cloud integrates cloud-based inspection and version recommendation provided by the eDesk to better support network administrators in O&M work.

10.3 DEVICE UPGRADE

Device vendors will keep tracking device and software issues, and resolve them by periodically releasing patches and new software versions. To ensure the stability of network devices, users are advised to keep the software version of devices up to date to prevent known network issues from occurring. In addition, update or upgrade also enables users to access or use new functions and features introduced after the previous installation.

10.3.1 Procedures

For the WAC upgrade mode, load the corresponding system software to a device through the server or with the web NMS function, and then set the device to restart upon the next software startup. Figure 10.3 shows how to upgrade the WAC using the web NMS function.

FIGURE 10.3   Web NMS upgrade.

The AP supports two upgrade modes. It can be upgraded either immediately or at a scheduled time.

1. WAC mode

   Upload the software package to the WAC through either the File Transfer Protocol (FTP), Secure File Transfer Protocol (SFTP), or Web to upgrade the APs. Each AP needs to download software from the WAC, and the software is downloaded to each AP one by one when there are a large number of APs.

2. FTP/SFTP mode

   Each AP downloads the software from the FTP/SFTP server, which will be sequentially downloaded to each AP if there are a large number of APs. The FTP/SFTP server's concurrency performance is much higher than that of the WAC, and therefore this method is recommended if there are a large number of APs.

Figure 10.4 illustrates how to use the web platform to upgrade APs in WAC or FTP/SFTP mode.

10.3.2 Viewing the Upgrade Status

When a WAC is upgraded, it cannot be connected to the network and its upgrade status cannot be viewed. When an AP is upgraded, its upgrade status can be viewed on the WAC. The AP's upgrade status will display

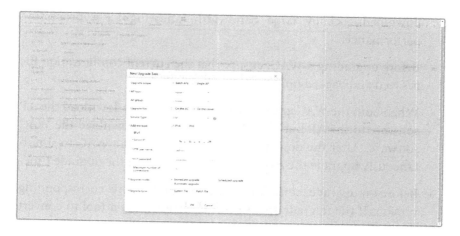

FIGURE 10.4   AP upgrade using the web system.

FIGURE 10.5   AP upgrade status display on the web platform.

either "waiting for download," "downloading," or "downloaded" as well as the upgrade status (succeed/failed and failure causes). Figure 10.5 shows the AP upgrade status display on the web platform.

10.3.3 Upgrade Precautions

When performing upgrades, the network administrator needs to consider the following issues:

- When upgrading a large number of APs, batch upgrade will prevent large-scale service interruption and heavy network load caused by software downloads from the server or WAC.

- If there are a large number of APs or the WAC and APs are deployed in different locations, such as in the headquarters and branches, downloading the software from the network closest to where the APs are located will ensure an optimal download speed. Do not download system software when AP traffic is heavy because software downloads occupy large bandwidth, affecting CPU processing performance.

- Before upgrading a WAC, ensure that the current configuration file is saved and backed up.

- When both the WAC and AP need upgrading, first download the AP software to the AP but do not install, and then download the WAC software. After the download is complete, reset the WAC to simultaneously upgrade the AP and WAC, minimizing the upgrade time.

- Before an upgrade, consult the upgrade guide of the corresponding version. If the upgrade spans several versions, use a tool to convert the configuration to avoid compatibility issues.

10.4 TROUBLESHOOTING

On a typical WLAN, network faults can be located by segment. Figure 10.6 illustrates the topology of a typical WLAN. Key links on the network are as follows:

1. STA-AP links, including the STA and wireless environment.

2. AP-switch links, including AP hardware, power supply, and system software.

3. WAC-switch links, including hardware connections, system software, and configurations between switches and WACs.

4. Switch-RADIUS/DHCP links, including the server's status, version information, and configurations.

When rectifying a fault, check all links in the faulty area.

- If the fault occurs on a single STA, check the link between the STA and the AP and then check whether the AP itself is faulty. Subsequently, check whether the switch or the WAC is faulty.

- If multiple STAs suffer from the same problem, check the other three links. Generally, check the links related to the WAC first.

Table 10.3 lists common WLAN faults.

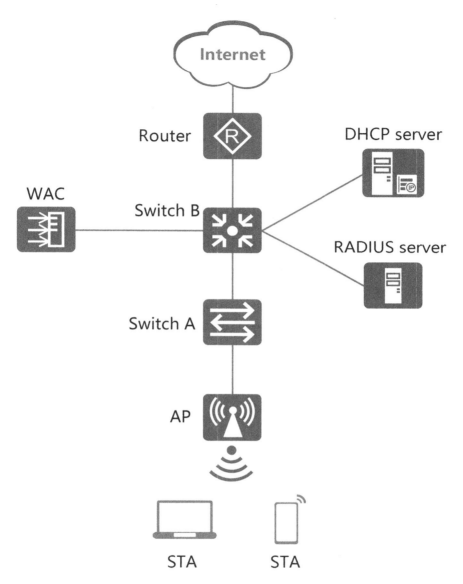

FIGURE 10.6   Typical WLAN networking.

The following uses the AP online failure as an example to describe how to troubleshoot common faults. For an AP to successfully go online, the links between the AP and the switch, between the switch and the WAC, and between the switch and the DHCP server must be normal. To locate the fault, perform the following steps:

TABLE 10.3   Common WLAN Faults

Category	Symptom
AP management	The AP fails to go online, disconnects unexpectedly, fails to upgrade, or fails to switch between fat and fit mode
User login	An STA fails to discover Wi-Fi signals, associate with the network, or be authenticated (using 802.1X authentication, built-in or external Portal authentication, or WeChat authentication)
User service performance	Low network speed, roaming failure, weak signal, or packet loss
WAC management	WAC upgrade failure, high CPU usage, and device login failure (Web/STelnet/Telnet)
WLAN services	The calibration does not take effect or delivers poor effect. APs cannot go online on the WDS or Mesh network or the link setup rate is low

Step 1: Check whether the WAC has received the online request from the AP, and whether the WAC has recorded any failed online requests from the AP. If it has, the link between the WAC and the AP is normal. In this case, consult the error information on the WAC before handling the fault.

- Some faults are caused by the configurations of the WAC or the AP. For example, the number of connected APs has reached the limit of WAC or license specifications. Other causes might be that the MAC address or device serial number of the AP is not configured on the WAC, the WAC and AP versions do not match, the AP is blacklisted, or the Datagram Transport Layer Security (DTLS) keys between the WAC and the AP are inconsistent.

- Some faults are caused by the configurations or connections in other links. For example, if an AP fails to obtain the DHCP address, perform step 2 to check the connection between the AP and the DHCP server. If no record of connection failure is displayed, go to step 3.

Step 2: Check the connection between the AP and the DHCP server by using the ping command. Specifically, isolate the link between the AP and the switch and then the link between the switch and DHCP server. It is important to also check the devices, cables, and related network and service configurations (interface and virtual local area network (VLAN) configurations, IP address configurations, DHCP address pool, and AP port mode).

Step 3: Check the connection between the WAC and the AP by using the ping command. Specifically, isolate the link between the AP and the switch and then the link between the switch and the WAC. Check the devices, cables, and network and service configurations, including the Control and Provisioning of Wireless Access Points (CAPWAP) source address, interface and VLAN configurations, IP address, and AP port mode.

If the fault persists, contact technical support personnel to collect device information for further analysis.

Fault diagnosis is a complex process that involves various troubleshooting methods. The preceding segment-by-segment troubleshooting procedure is recommended. Table 10.4 lists the capabilities provided by mainstream devices and common references for troubleshooting.

TABLE 10.4   Capabilities Provided by Mainstream Devices and References for Fault Diagnosis

O&M Capability	O&M Function	Method
Fault detection capability	Fault alarm monitoring	The device continuously monitors important indicators, such as an AP's offline status, user access limit, and high CPU usage. Once an issue is detected, the device reports an alarm. The network administrator can view alarm information on the monitoring platform or in device logs
	Metric monitoring	Important WLAN indicators are described in Section 10.1
	Fault visualization	Important faults, such as irregular AP status and air interface quality deterioration, can be summarized and displayed on the monitoring platform. For example, the web system provides the offline rate and the channel usage distribution of APs
Fault collection capability	Syslog	User logs can be reported to the log server deployed on the local network through Syslog
	Log collection	The user logs, operation logs, and diagnosis logs saved on the device can be downloaded for problem analysis
	Indicator collection	As described in Section 10.1, the device's routine operation indicators can be collected by a network management device via SNMP or Telemetry
	One-click information collection	The device provides the one-click information collection capability to traverse main monitoring and diagnosis commands to collect important information about each module for subsequent fault locating

(Continued)

TABLE 10.4 (*Continued*)   Capabilities Provided by Mainstream Devices and References for Fault Diagnosis

O&M Capability	O&M Function	Method
Fault diagnosis capability	Command line	You can remotely log in to the command line interface (CLI) of a device to run commands
	Equipment information analysis	Based on the exported log files, indicators, and information collected in one-click mode, perform module-level analysis focused on the fault occurrence period
	Packet capturing	Use an auxiliary packet capture tool to obtain packets on the related network segment or device to facilitate fault locating
Fault diagnosis-related case reference	Maintenance documentation	The technical support website of device vendors provides many maintenance documents that describe how to troubleshoot WLAN faults and how to demarcate common faults

10.5 INTELLIGENT O&M

Currently, WLAN O&M mainly focuses on devices. However, user service provisioning is conducted on an end-to-end basis, and any network fault may cause user experience to deteriorate. In addition, network monitoring and information displays are utilized to passively indicate the network status. Administrators rely on user reporting to detect network faults, and this mechanism cannot be used to display network-wide service indicators or automatically detect potential network problems. Given this, users are likely to detect network problems before administrators do.

Ideally, network O&M should be upgraded from device-centric to user experience-centric AI-based intelligent O&M, improving user experience based on the prediction and intelligence of O&M systems. In the next-generation intent-driven campus network, the software-defined networking (SDN) controller provides an intelligent and automatic proactive network analysis system that integrates both big data collection and AI computing capabilities. Furthermore, the SDN controller predicts faults and adjusts the network in advance to reduce the fault rate, as illustrated in Figure 10.7.

Utilizing powerful AI analysis capability and data storage capabilities on servers, the SDN controller can display indicators for both the overall network and user experience, identify and locate problems, as well as provide auxiliary functions such as proactive notification and O&M report

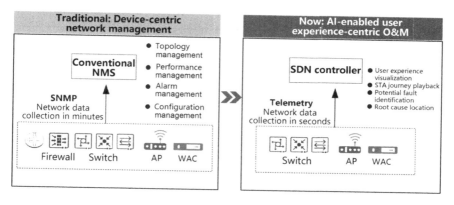

FIGURE 10.7   Intelligent O&M.

generation. This enables network administrators to manage networks in a comprehensive and systematic manner, while also simplifying network O&M. See Table 10.5.

10.5.1 User Experience Visibility

To achieve experience visibility, sufficient data must be available and obtained quickly to ensure accurate analysis results. Therefore, data collection is vital, and it involves WLAN quality KPIs being measured from APs, radio frequencies, and STAs. Table 10.6 lists data to be collected and the minimum sampling duration. After data collection is complete, the system can proactively identify air interface performance and connection issues—such as weak signal coverage, severe interference, and high channel usage—based on AI algorithms, correlation analysis, and exception modes.

For example, in traditional O&M, O&M personnel cannot perceive the most common determinants of poor user experience on WLANs, and they therefore cannot optimize networks promptly as problems can only be rectified after fault reports are received. When locating a fault, professional engineers must visit the site to simulate faults or repeatedly recreate them. With intelligent O&M, the system can automatically analyze the quality of network-wide user experience and provide detailed experience data of poor-QoE users. When a poor-QoE user is detected, O&M personnel can quickly resolve the problem by viewing the occurrence time and cause of the poor-QoE problem, as illustrated in Figure 10.8.

TABLE 10.5   Intelligent O&M Capabilities

Function	Description
User experience visualization	**Any moment**: Using the telemetry technology, network KPI data is dynamically captured in seconds in real time, allowing processes to be restored and faults to be traced **Any user**: Multidimensional data collection helps display users' network profiles of all users in real time and visualize network experience throughout the process, including STAs, access time, APs, user experience, and fault **Any application**: This includes real-time voice and video application experience awareness, as well as visualized audio and video application processes
Fault identification and proactive prediction	**Automatic fault identification**: Big data and AI technologies are used to automatically identify issues regarding connectivity, air interface performance, roaming, device, and application issues. This improves the identification rate of potential issues **Proactive fault prediction**: Historical data is obtained through machine learning to dynamically generate a baseline, which is then compared and analyzed with real-time data to predict possible faults
Fault locating and root cause analysis	**Quick fault locating**: This is used to intelligently identify issue patterns and impact scopes based on the network O&M expert system and various AI algorithms, enabling administrators to accurately locate issues **Intelligent root cause analysis**: Based on the big data platform, possible causes are analyzed and possible solutions are provided. For example, to resolve optimization algorithm problems, these algorithms can perform automatic optimization based on the network-wide terminal profile
Proactive notification	**Proactive notification**: Emails or short message service (SMS) messages are sent to notify users of major problems **O&M report**: O&M reports and summaries are pushed at scheduled intervals

TABLE 10.6   WLAN Data

Measurement Object	Main Measurement Indicator	Collection Device	Minimum Sampling Period (s)
AP	CPU and memory usage, and number of online users	AP	10
Radio frequency	Number of online users, channel usage, noise, traffic, backpressure queue, interference rate, and power	AP	10
STA	RSSI, negotiated rate, packet loss rate, delay, DHCP, and 802.1X authentication	AP	10

Poor-QoE user

XXXXXX

XXXX

User journey

 According to analysis, user x is a poor-QoE user. The total online duration is 10 hours, 15 minutes, and 30 seconds, and the poor-quality duration is 3 hours and 30 minutes. The poor-quality duration accounts for 34.12%

The following causes are located:

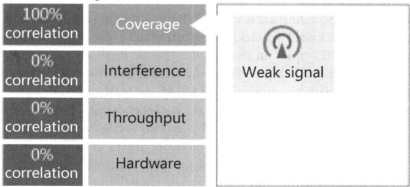

FIGURE 10.8   Poor-QoE user analysis.

10.5.2 Fault Identification and Proactive Prediction

In traditional network O&M, problems can only be solved after they occur. If faults can be predicted, potential problems can be detected in advance. This enables the implementation of necessary measures, such as network hardening or rectification, to mitigate the impact on services.

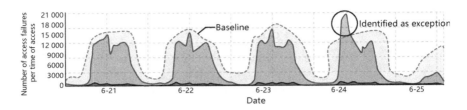

FIGURE 10.9   Fault baseline and exception detection.

Based on big data analysis, intelligent O&M fault identification and proactive prediction can be used to predict certain network faults and provide warnings. For example, users may experience network access failures that are not caused by network faults. As illustrated in Figure 10.9, the SDN controller generates a baseline based on historical big data training. Within this baseline, failures and exceptions are considered to be terminal behaviors. The system automatically identifies exceptions only when they are beyond the baseline range, then it identifies patterns and root causes, as well as faults promptly, and notifies O&M personnel, enabling fault handling before users are aware.

10.5.3 Fault Locating and Root Cause Analysis

Network O&M personnel are responsible for maintaining networks. When a fault occurs, they need to quickly identify the fault cause, rectify the fault, and minimize the impact on services. Traditional methods for locating a fault are difficult due to high reliance on manual analysis of massive data and personal experience. In intelligent O&M, the SDN controller can use protocol tracing to graphically display the packet exchange process when a fault occurs, enabling O&M personnel to quickly locate the fault.

For example, if a user encounters network access difficulties or failures, the protocol tracing function of the SDN controller can visualize the entire process in the three phases (association, authentication, and DHCP) of user access. Subsequently, by analyzing the result and duration of each protocol interaction phase, the SDN controller can quickly determine where user access errors exist, and implement precise fault location, as illustrated in Figure 10.10.

10.5.4 Network O&M Mode Transformed by Intelligent O&M

With the powerful SDN controller and mobile apps, imagine how convenient it will be for network administrators to manage networks in the near future.

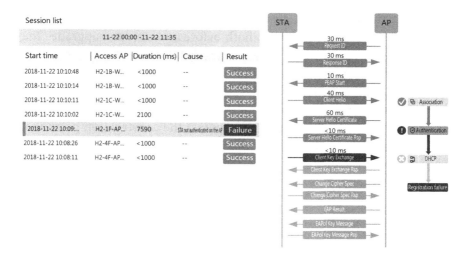

Session list				
11-22 00:00 -11-22 11:35				
Start time	Access AP	Duration (ms)	Cause	Result
2018-11-22 10:10:48	H2-1B-W...	<1000	--	Success
2018-11-22 10:10:14	H2-1B-W...	<1000	--	Success
2018-11-22 10:10:11	H2-1C-W...	<1000	--	Success
2018-11-22 10:10:02	H2-1C-W...	2100	--	Success
2018-11-22 10:09:...	H2-1F-AP...	7590	STA not authenticated on the AP	Failure
2018-11-22 10:08:26	H2-4F-AP...	<1000	--	Success
2018-11-22 10:08:11	H2-4F-AP...	<1000	--	Success

FIGURE 10.10   Protocol tracing.

Suppose you are a network administrator. On Wednesday, you open the controller dashboard to check for normal indicators and faults. You can perform these tasks on your mobile app if you are, for example, in a meeting.

At 10:00 a.m., the platform pushes a new device version, which you check for changes and discover that they can solve a previous issue. As such, you set a scheduled upgrade and plan to perform the upgrade at 10:00 p.m. on Friday, as there will be few network users at night and during weekend.

At 3:00 p.m., the platform indicates that a large number of users failed to access the network, and you receive an SMS notification simultaneously. You click the link to check the details page and discover that the DHCP address could not be obtained. You then log in to the DHCP server and realize that the same problem previously occurred. The new server version was not released but a solution was provided. You restore the server by following instructions in the solution. After the operation is complete, you receive a call from a colleague, explain the situation, and notify them that the fault has been rectified and the network will recover soon. Then you send a group message through the internal communication platform to notify colleagues of the situation.

At 6:00 p.m. on Friday, you leave work on time. At 10:00 p.m., your mobile app sends a message to notify you that the upgrade has started. At

10:30 p.m., the app sends a message to notify you of upgrade completion and sends network O&M reports both before and after the upgrade. The reports indicate that the APs are working properly.

On Sunday morning, the construction team starts to reconstruct the equipment room as scheduled. At 11:00 a.m., you receive a major alarm indicating that 50 APs went offline because the switch interfaces connected to the APs are abnormal. You contact the construction personnel to check the equipment room status and realize that the power supply of the switch was accidentally turned off during construction. One minute after the power supply is restored, the network recovers.

Acronyms and Abbreviations

16QAM	16 Quadrature Amplitude Modulation
AAA	Authentication, Authorization, and Accounting
AC	Access Category
ACK	Acknowledgment
AES	Advanced Encryption Standard
AGC	Automatic Gain Control
AGV	Automated Guided Vehicle
AI	Artificial Intelligence
AID	Association ID
AIFS	Arbitration Interframe Space
AIFSN	Arbitration Interframe Spacing Number
ALSNR	Alien Limited Signal-to-Noise Ratio
AMC	Adaptive Modulation and Coding
A-MPDU	Aggregate Media Access Control Protocol Data Unit
A-MSDU	Aggregate MAC Service Data Unit
AP	Access Point
API	Application Programming Interface
AR	Augmented Reality
ARP	Address Resolution Protocol
ASK	Amplitude Shift Keying
AWG	American Wire Gauge
BA	Block Acknowledgment
BAR	Block Acknowledgment Request
BCC	Binary Convolutional Coding
BFRP	Beamforming Report Poll

BIM	Building Information Model
BLE	Bluetooth Low Energy
BPSK	Binary Phase Shift Keying
BSA	Basic Service Area
BSR	Buffer Status Report
BSS	Basic Service Set
BSSID	Basic Service Set Identifier
BTF	Basic Trigger Frame
BTM	BSS Transition Management
BYOD	Bring Your Own Device
C/S	Client/Server
CAPEX	Capital Expenditure
CAPWAP	Control and Provisioning of Wireless Access Points
CC	Content Channel
CCA	Clear Channel Assessment
CCB	Contiguous Channel Bonding
CCK	Complementary Code Keying
CCMP	Counter Mode with CBC-MAC Protocol
CES	Consumer Electronics Show
CFP	Contention-Free Period
CPE	Customer-Premises Equipment
CPU	Central Processing Unit
CQI	Channel Quality Indicator
CR	Channel Reservation
CRC	Cyclic Redundancy Check
CS	Carrier Sense
CSI	Channel State Information
CSMA/CA	Carrier Sense Multiple Access with Collision Avoidance
CSMA/CD	Carrier Sense Multiple Access with Collision Detection
CT	Computerized Tomography
CTS	Clear to Send
DBS	Dynamic Bandwidth Selection
DCA	Dynamic Channel Allocation
DCF	Distributed Coordination Function
DCM	Dual-Carrier Modulation
DFA	Dynamic Frequency Assignment

DFBS	Dynamic Frequency Band Selection
DHCP	Dynamic Host Configuration Protocol
DIFS	DCF Interframe Space
DMZ	Demilitarized Zone
DNS	Domain Name System
DS	Distribution System
DSCP	Differentiated Services Code Point
DSSS	Direct Sequence Spread Spectrum
DTLS	Datagram Transport Layer Security
E2E	End-to-End
EAP-PEAP	Extensible Authentication Protocol-Protected Extensible Authentication Protocol
ECC	Envelope Correlation Coefficient
ED	Energy Detect
EDCA	Enhanced Distributed Channel Access
EHT	Extremely High Throughput
EIA	Electronic Industry Association
EMR	Electronic Medical Record
ENP	Ethernet Network Processor
ERP	Enterprise Resource Planning
ESL	Electronic Shelf Label
ESPRIT	Estimation of Signal Parameters via Rotational Invariance Technique
ESS	Extended Service Set
ESSID	ESS Identifier
FAP	Foreign AP
FCC	Federal Communications Commission
FCS	Frame Check Sequence
FDMA	Frequency Division Multiple Access
FFT	Fast Fourier Transformation
FHSS	Frequency Hopping Spread Spectrum
FIFO	First-In-First-Out
FSK	Frequency Shift Keying
FT	Fast Basic Service Set Transition
FTM	Fine Timing Measurement
FTP	File Transfer Protocol
FWAC	Foreign WAC
GFSK	Gaussian Frequency Shift Keying

GI	Guard Interval
HAP	Home AP
HE ER SU PPDU	High Efficiency Extended Range Single-User PPDU
HE MU PPDU	High Efficiency Multiuser PPDU
HE SU PPDU	High Efficiency Singleuser PPDU
HE TB PPDU	High Efficiency Trigger-Based PPDU
HE-LTF	High Efficiency Long Training Field
HE-SIG-A	High Efficiency Signal Field A
HE-SIG-B	High Efficiency Signal Field B
HE-STF	High Efficiency Short Training Field
HEW	High Efficiency WLAN
HF	High Frequency
HSB	Hot Standby
HT	High Throughput
HT-LTF	High Throughput Long Training Field
HT-STF	High Throughput Short Training Field
HTTP	Hypertext Transfer Protocol
HWAC	Home WAC
I/Q	In-phase/Quadrature
IDEA	Information Digitalization and Experience Assurance
IEEE	Institute of Electrical and Electronics Engineers
IFFT	Inverse Fast Fourier Transformation
IFS	Interframe Space
IP	Internet Protocol
IPsec	Internet Protocol Security
ISD	Inter-Site Distance
ISI	Intersymbol Interference
ISM	Industrial, Scientific, and Medical
IV	Initialization Vector
KPI	Key Performance Indicator
KRACK	Key Installation Attack
LAN	Local Area Network
LBS	Location Based Service
LDPC	Low-Density Parity-Check
LF	Low Frequency
LLC	Logical Link Control
L-LTF	Legacy Long Training Field
LOS	Line-of-Sight

L-SIG	Legacy Signal Field
L-STF	Legacy Short Training Field
M2M	Machine-to-Machine
MAC	Media Access Control
MAN	Metropolitan Area Network
MBA	Multistation Block Acknowledgment
MCS	Modulation and Coding Scheme
MDID	Mobility Domain Identifier
MIB	Management Information Base
MIC	Message Integrity Check
MIMO	Multiple-Input Multiple-Output
MISO	Multiple-Input Single-Output
MITM	Man-in-the-Middle
MMPDU	MAC Management Protocol Data Unit
MPDU	MAC Protocol Data Unit
MRC	Maximum Ratio Combining
MRI	Magnetic Resonance Imaging
MSDU	MAC Service Data Unit
MU-BAR	Multiuser Block Acknowledge Request
MU-MIMO	Multiuser Multiple-Input Multiple-Output
MU-RTS	Multiuser Request to Send
MUSIC	Machine Utilization Statistical Information Collection
N_STA	Number of Scheduled STAs
NAV	Network Allocation Vector
NCB	Noncontiguous Channel Bonding
NDP	Null Data Packet
NFRP	NDP Feedback Report Poll
NLOS	Non-Line-of-Sight
NMS	Network Management System
OBO	OFDMA Backoff
OBSS	Overlapping Basic Service Set
OBSS-PD	Overlapping Basic Service Set–Packet Detect
OCW	OFDMA Contention Window
OFDM	Orthogonal Frequency Division Multiplexing
OFDMA	Orthogonal Frequency Division Multiple Access
OMI	Operating Mode Indication
OPEX	Operating Expense
OSA	Open System Authentication

OUI	Organizational Unique Identifier
PA	Power Amplifier
PACS	Picture Archiving and Communication System
PAPR	Peak to Average Power Ratio
PB	Power Boosting
PCF	Point Coordination Function
PC	Personal Computer
PDA	Personal Digital Assistant
PE	Packet Extension
PER	Packet Error Rate
PIFA	Planar Inverted F Antenna
PIFS	PCF Interframe Space
PLCP	Physical Layer Convergence Procedure
PMD	Physical Medium Dependent
PMK	Pairwise Master Key
PMKID	PMK Identifier
PMKSA	Pairwise Master Key Security Association
PPDU	PLCP Protocol Data Unit
PPSK	Private Pre-Shared Key
PSDU	PLCP Service Data Unit
PSK	Phase-Shift Keying
PSK	Pre-Shared Key
PTK	Pairwise Transient Key
QAM	Quadrature Amplitude Modulation
QBPSK	Quadrature Binary Phase Shift Keying
QPSK	Quadrature Phase Shift Keying
RADIUS	Remote Authentication Dial-In User Service
RC4	Rivest Cipher 4
RDG	Reverse Direction Grant
RF	Radio Frequency
RFID	Radio Frequency Identification
RL-SIG	Repeated Legacy Signal Field
RPL	Received Power Level
RR	Resource Request
RSSI	Received Signal Strength Indicator
RTP	Real-time Transport Protocol
RTS	Request To Send
RTS/CTS	Request to Send/Clear to Send

RTT	Round Trip Time
RU	Remote Unit
RU	Resource Unit
SAC	Smart Application Control
SD	Signal Detect
SDMA	Space Division Multiple Access
SDN	Software-Defined Networking
SFN	Single Frequency Network
SFTP	Secure File Transfer Protocol
SHF	Super High Frequency
SIFS	Short Interframe Space
SIMO	Single-Input Multiple-Output
SINR	Signal-to-Interference-plus-Noise Ratio
SIP	Session Initiation Protocol
SISO	Single-Input Single-Output
SKA	Shared Key Authentication
SMS	Short Message Service
SNMP	Simple Network Management Protocol
SNR	Signal-to-Noise Ratio
SOHO	Small Office/Home Office
SRG	Spatial Reuse Group
SRP	Spatial Reuse Parameter
SSID	Similar Service Set Identifier
SSL	Secure Sockets Layer
SU-MIMO	Single-User MIMO
SWR	Standing Wave Ratio
TCP	Transmission Control Protocol
TDD	Time Division Duplex
TKIP	Temporal Key Integrity Protocol
TPC	Transmit Power Control
TWT	Target Wake-up Time
TXOP	Transmit Opportunity
UHF	Ultra High Frequency
UIF	User Information Field
ULA	Uniform Linear Array
UNB	Ultra Narrow Band
UNII	Unlicensed National Information Infrastructure
UORA	Uplink OFDMA-based Random Access

UP	User Preference
URL	Uniform Resource Locator
USB	Universal Serial Bus
UWB	Ultra-Wide Band
VAP	Virtual Access Point
VHT	Very High Throughput
VHT-LTF	Very High Throughput Long Training Field
VOD	Video On Demand
VPN	Virtual Private Network
VR	Virtual Reality
VRRP	Virtual Router Redundancy Protocol
WAC	Wireless Access Controller
WAI	WLAN Authentication Infrastructure
WAPI	WLAN Authentication and Privacy Infrastructure
WDS	Wireless Distribution System
WEP	Wired Equivalent Privacy
WIDS	Wireless Intrusion Detection System
WIPS	Wireless Intrusion Prevention System
WLAN	Wireless Local Area Network
WMM	Wi-Fi Multimedia
WPA	Wi-Fi Protected Access
WPA2	Wi-Fi Protected Access 2
WPA3	Wi-Fi Protected Access 3
WPI	WLAN Privacy Infrastructure
ZC	ZigBee Coordinator
ZED	ZigBee End Device
ZR	ZigBee Router

Bibliography

For technical details and specific parameters, see the recommended books and related documents on the IEEE Standardized Association official website.

1. Li Y B, Li Y C, Liu L, et al. Non-contiguous channel bonding in 11ax[EB/OL]. (2016-01-17).
2. Fischer M, Seok Y H. Disallowed sub channels[EB/OL]. (2018-04-16).
3. Ghosh C, Stacey R, Perahia E, et al. Random access with trigger frames using OFDMA[EB/OL]. (2015-05-12).
4. Guo J Y C, Yang D X, Li Y B. Comment resolution on trigger frame for random access[EB/OL]. (2017-05-09).
5. Ghosh C, Stacey R, Perahia E, et al. UL OFDMA-based random access procedure[EB/OL]. (2015-09-14).
6. Kim J S, Mujtaba A, Li G Q, et al. 20 MHz-only device in 11ax [EB/OL]. (2016-07-25).
7. Kim J S, Mujtaba A, Li G Q, et al. RU restriction of 20MHz operating devices in OFDMA[EB/OL]. (2016-07-25).
8. Li G Q, Kneck J, Hartman C, et al. CIDs related to 20MHz-only STAs operating on non-primary 20 MHz channels[EB/OL]. (2017-03-13).
9. Seok Y H, Wang C C, Yee J, et al. LB230 CR 20MHz only STA on secondary channel[EB/OL]. (2018-03-15).
10. Gao F, Li P, Yang W, et al. *HCNA-WLAN Learner's Guide* [M]. Beijing: People's Posts and Telecommunications Press, 2015.

Printed in the United States
by Baker & Taylor Publisher Services